Contextual Psychology

Integrating Mindfulness-Based Approaches Into Effective Therapy

————————————

Dr. Richard K. Nongard

Contextual Psychology:
Integrating Mindfulness-Based Approaches Into Effective Therapy
Dr. Richard K. Nongard

Edited by James Hazlerig (www.HarmonyHypnosis.net)
Cover graphics by Dustin Tilly-Sturgeon (www.VividGraphicDesign.ca)

First Printing: March 2014

ISBN: 978-1-304-94913-4

PeachTree Professional Education, Inc.
7107 S. Yale, Ste 370
Tulsa, OK 74136
(918) 236-6116
www.SubliminalScience.com
www.FastCEUs.com

About the Author

Dr. Richard Nongard is a licensed marriage and family therapist with a Master's Degree in Counseling from Liberty University and a Doctorate in Transformational Leadership (Cultural Transformation) from Bakke Graduate University.

He has written a number of different textbooks on counseling and clinical hypnosis. Directly related to the content in this book is his book *Medical Meditation*, which has been adapted in a number of different settings to help both pre-surgical and post-surgical clients experience health and wellness throughout their operations. He also wrote a book on Contextual Hypnotherapy, which integrates many of the techniques discussed in this book into the practice of clinical hypnotherapy.

He currently has a private practice in Scottsdale, Arizona, where he sees clients on a regular basis. In the past, his work has included both inpatient and outpatient psychiatric and substance abuse settings, where he's worked with a wide variety of clients over the years.

Table of Contents

Introduction vii

CHAPTER ONE
What Is Contextual Psychology?.................................... 1

CHAPTER TWO
Relational Frame Theory 11

CHAPTER THREE
Challenging the Idea of Past-Tense Therapy..........................23

CHAPTER FOUR
Mindfulness as a Success Foundation.................................29

CHAPTER FIVE
ACT Therapy Solutions...39

CHAPTER SIX
Positive Psychology...59

CHAPTER SEVEN
Solution-Focused Brief Therapy 75

CHAPTER EIGHT:
Mindfulness-Based Stress Reduction (MBSR) And Other
Mindfulness-Based Therapies...89

CHAPTER NINE
Dialectical Behavioral Therapy....................................105

CHAPTER TEN
DBT Skills Training..115

CHAPTER ELEVEN
Metaphor and Story 137

CHAPTER TWELVE
Case Aplication and Structure of Therapy 151

Bibliography...161

Introduction

The purpose of this book is very simple: I want to provide information that's going to help you to re-ignite your passion for your work as a therapist, counselor, social worker, or really as any member of the helping professions.

Remember when you first graduated, you walked down the aisle and you thought about that first job where you were going to help people get well. It was an amazing feeling, wasn't it? I remember how excited I was as a new graduate with my Master's Degree, when I had finished my practicum and had already secured my first full time job as a therapist. Some of those early clients that I worked with, even though that was more than a quarter century ago, are clients that I still think about today.

I enjoyed feeling like I had an impact on their life, that the work that I was doing was meaningful, and that I was helping them to make substantial changes. Certainly, I'm not the only one with those experiences. I bet you can think back to a time when you were excited and passionate about the work that you did. Maybe it was particular setting where you had the opportunity to work with clients that were close to your heart, or maybe it was in private practice when you had those clients who were self-motivated and actually utilized the resources you taught them to make significant and positive changes. Chances are pretty good that you're still passionate about the work that you do.

We all got in this field for two primary reasons, I like to say. One: we didn't have to take math to get the degree. Two: ultimately we really like helping people make change. In the era of managed care though, in the era of limited time and limited resources, in the era of significant changes in the way that we deliver services year after year of doing the work of a counselor or a social worker, registered nurse in a psychiatric setting or really any other member of the helping profession, you may have found that going to the office, going to the clinic, going to the community service center, wherever it is that you work has lost a little bit of its luster.

I think sometimes this is in part due to the fact that most of us have been doing the same thing that we were taught to during graduate school throughout our career as professional counselors, social workers, and members of the helping profession. Therapy school claims it teaches us ideas, but what it actually teaches us are *old* ideas. Because of this, as we do our work day after day and work with other counselors who are more recent graduates, we can feel inadequate or less effective because of the newer knowledge they have that we didn't have the opportunity to learn.

This book is about teaching you new ideas, and when you learn new ideas, you'll learn to re-ignite that passion you feel about the social work and counseling professions. Now, graduate school for therapy teaches us a lot, and this is a good thing, but it is important to note that all schools teach old ideas. For example, I have a son who is in his senior year of high school, and he's making decisions about college. He's looking forward to all of the things he's going to learn in his field, he says. He and I have had discussions about which school he's going to choose, what he's going to study, and of course what the cost and constraints and the geography of what those decisions mean. At times we haven't always agreed on every issue related to the subject, and he of course has said, "But I'm going to learn all these things that I'm going to be able to use in my job!"

My answer to him was "Well, actually what you're going to do in school as an undergraduate is, you're going to learn how to learn, because when you get your first job they're going to re-train you in today's methods. The reason why they're going to re-train you in today's methods is because everything you learned as an undergraduate was actually old ideas." This is the nature of education. We only learn what has already been learned. Unless you're working in or studying at a graduate school and doing specific research in the new modalities, chances are pretty good you're actually learning old content.

An easier comparison would be to consider the print newspaper. Print newspapers contain only yesterday's stories with the same technology for the past fifty years, and they are physically inhibited from having up-to-the-minute news stories. The Internet, however, has the capability of breaking news seconds after it happens, and having ever-evolving coverage on a particular topic with all of the latest techniques for reporting. In this analogy, your undergraduate work is the print news, and your actual work experience is the Internet.

As we enter into the professions where we work, the business of life in the inadequacy of most state requirements related to continuing education mean that many of us never actually have the chance or the opportunity to learn the newest and most effective techniques of therapy. If you're like me when you graduated more than a century ago, your practice probably has been based on what you learned in the past and what feels like it's helping your clients.

In the era of managed care, in the current era where we're working though different standard measures of therapeutic success, there is a lot to constantly learn. We're in the era now of evidence-based treatment protocols. If you're a more recent graduate of a counselor program or you've been particularly careful in your selection of continuing education courses, you might have an edge from an academic perspective, but chances are you learned about therapy in many of these courses rather than actually having to do therapy.

This book is both about new therapeutic approaches and how to actually enact them with your clients. First, I'm going to focus on an overview of Contextual Psychology, but in each successive chapter I'll not only share a theoretical or a philosophical idea of a particular therapeutic approach, I'll also describe the experiences with my clients in my office. I'm also going to demonstrate with you in an interactive way the various processes that are of value to you.

This book will help you to re-ignite your passion by directing you to do several things. First of all, it will help you to understand the new philosophies of counseling and therapy based on current research. A lot of therapists like myself practice day in and day out doing what we were taught in graduate school, and yet there's been so much more in the last 25 years which has been empirically validated as efficient treatment for a wide variety of different behavioral psychological disorders.

This course is going to overview those philosophies that are based on current research in contemporary university settings, but it's not going to be all academic. I'm going to teach you how to take that research and actually apply it to those clients who are in your office. This book is going to help you to develop a map for navigating tough or unfamiliar therapeutic settings.

Every now and then I get a call from a client and the presenting problem from either that client or from the family is something that I really don't have a lot familiarity with. It's new. It's unique. It's a different type of presenting problem than I've worked with in the past. I really enjoy these types of cases because it gives me the opportunity to sort of get out of a rut—to do something new and apply my skills with a totally different dynamic or a totally different set of presenting problems. Sometimes, however, in these unfamiliar therapeutic settings or with a particularly tough client I think to myself (as I'm sure you do also), "What should I be doing in this session?"

By understanding Contextual Psychology, you will have the skills necessary to steer through these situations with confidence and grace. You will know where to start, what to do in that first session with that client, what can be done over the next six to eight weeks to assist that client. You're going to feel confident in your ability to help those people who are on your caseload. This book is going to help you re-ignite your passion by helping you to develop specific skills and processes for effective therapy.

I remember when I was a new counselor, back in the day when our practice were only 300 hours when we needed only 2,000 hours before we actually could qualify for licensure. For most of us those supervision hours really came in the form of doing full time work on your own and meeting with the supervisor every couple of weeks to basically have them sign off on your documents. I know that in the last twenty-five years requirements have become much more rigid than they were back in the day, but to a large extent I think many of us still have done most of our work as a trial and error on our own, without working hand-in-hand with more experienced therapists.

As a result, we've developed a repertoire of skills and processes that we use in therapy, but because we're probably not working with another therapist and watching the work that they do, we're not expanding our repertoire of specific skills and processes. Over the course of this book, I'm going to share with you processes I use for effective therapy. You will know what technique to use, for example, to help somebody overcome a fear of flying, or to help somebody let go of depression that has been holding them back from success.

You will learn specific skills and processes, like things to do twenty minutes into the therapy session when you run into certain issues that your clients wish to address. This book is going to jumpstart your practice by helping you to create change in your methods. In this business, change is the only constant, and keeping on top of the game is the most important thing you can do to keep your passion alive and your clients happy.

When clients ask "How do you know this will be helpful to me?" you will have adapted a set of methods that are empirically based and that research shows have a greater efficacy in helping a larger number of people to experience the success for which they came to therapy in the first place.

The methods that you're going to be learning in this book include a variety of different approaches to trauma therapies. Almost every client who comes to our offices has experienced or is in the middle of experiencing significant trauma; fortunately for all, over the last twenty to twenty-five years, there has been significant amounts of research that has really shed light on the best approaches for helping people resolve trauma. This book is going to focus on giving you some of those strategies based on empirical research that can help you to help those who are experiencing trauma.

The methods of this book include a set of skills all with roots in what is referred to as Contextual Psychology. This includes those most closely associated with Contextual Psychology: Relational Frame Theory (RFT), Dialectical Behavioral Therapy (DBT), and Acceptance and Commitment Therapy (ACT).

Because of my experience in working with personality disorder clients, I've spent a great deal of time working as a therapist within the context of substance abuse and the criminal justice system. As a result of my work in the field, I have considered myself a student of Dialectical Behavioral Therapy since the early 1990s. Dialectical Behavioral Therapy, published by Marsha Linehan in 1993, called *Skills Training Manual for the Treatment of Borderline Personality Disorder*, is really one of the first approaches that we could put into the "basket" of Contextual Psychology.

That book really started a revolution in the treatment of personality disordered individuals, but the great thing about the last twenty-five to thirty years is that those techniques have been applied far beyond the original context of borderline personality disorder to a wide variety of clients, from self-mutilators to clients with eating disorders. This process has helped people who have a variety of different presenting problems experience success in a way they simply couldn't before.

This book is going to teach you how to use metaphor in therapy and changework. Contextual psychologists almost always have one thing in common, and that is the ability to recognize the value of language and language patterns in change work. You too are going to have that ability, and you are going to learn how to use metaphor in a way that's meaningful to help you communicate more effectively with your clients.

This course will also teach you the principles of Mindfulness-Based Stress Reduction (MBSR). Jon Kabat-Zinn at the University of Massachusetts in his pain management program almost 30 years ago began studying the efficacy of Mindfulness-based approaches in psychotherapy in a secular and non-religious setting. I'm going to share with you the ideas of Mindfulness-Based Stress Reduction and how they can be useful to the clients whom you work with in your office.

Mindfulness is a strategy that is really at the core of many of the different approaches of Contextual Psychology. I've always been a big fan of Solution-Focused Brief Therapy (SFBT), which is another evidence-based treatment protocol that has demonstrated efficacy with a wide range of clients and when we can utilize the principles of Solution-Focused Brief Therapy like the Miracle Question to help us determine what the most effective outcomes and treatment are going to be, we can be even more effective in therapy.

Although I've been familiar with Positive Psychology as a therapeutic technique for a long time, it wasn't until my doctoral studies at Bakke Graduate University that I began to use a process called Appreciative Inquiry in organizational management and consulting. The organizational or management approach to Appreciative Inquiry is really an adaptation or an outgrowth of the ideas of Positive Psychology that we can utilize with the clients on our caseload.

Sitting on my desk right now is one of my favorite resources: the

Oxford Handbook of Methods in Positive Psychology. It was edited by Anthony Ong and Manfred von Dulmen. It's a tremendous resource that I keep on my desk, and it's got dog ears, underlines, and yellow stickies all over it. I'm going to be sharing with you the things that I've yellow stickied, the things that I've dog eared, and the things that I've underlined so that when you have clients in your office at 4:15, you will know what to do with them, based on empirically validated approaches.

The ideas of Positive Psychology are really completely different than our traditional diagnostic or pathological model of psychology. Of course, one of the first evidence based treatment protocols was Cognitive Behavioral Therapy (CBT), associated with Albert Ellis and Aaron Beck.

CBT is the therapeutic approach that I spent the majority of my time learning in graduate school in the late 1980s. The reality is though that cognitive therapies have made significant improvements and changes over the last 20 to 25 years. Some of my favorite research comes from University of Toronto and the research into Mindfulness based cognitive therapies. I'm also going to talk about multi-component cognitive therapies, and you'll going to learn how to take your Cognitive Behavioral Therapy to the next level.

I'm also a big fan of Experiential Theater as Don Gibbons, psychologist from New Jersey, a former president of the New Jersey Chapter of the APA, calls it. I'm going to be sharing with you some of the techniques of experiential therapy where we don't just interview our client, reflect back what we heard them say and talk to our client, but instead we create an opportunity for them to take action in our office through experiential processes that they can practice between now and their next session.

I'm a firm believer in therapeutic homework. All of my clients know what they should be doing between now and next week, and experiential therapy gives us a great toolbox to draw from in helping our clients through the larger umbrella or basket of Contextual Psychology.

All of the material in this course is based on empirically supported treatment approaches. It's really important to know and use what psychological research has shown us to be most effective. This is what is required in the era of managed care. More importantly, using evidence-based approaches of course helps you at the end of the day to feel even better about the works that you do because you know there's a greater likelihood that your clients are going to experience the results they came in for. You'll be learning what really is the subject of current research and you'll be learning the specific processes and methods of using these strategies with clients in your office.

Now, let's move from future pacing what you are going to learn and actually begin learning right now. In fact, one of the themes of this book is the only time that you actually have is right now. Yesterday is gone. Tomorrow is not here. And so living in the present is really one of the key characteristics to helping our clients experience success in Contextual Psychology.

CHAPTER ONE

What Is Contextual Psychology?

All of the approaches that I just mentioned in the introduction fit underneath the banner of Contextual Psychology. Broadly, Contextual Psychology is the application of therapeutic philosophy that interprets an event as an ongoing act, inseparable either from its current or its historical context and in which a radically functional approach to truth and meaning is adapted.[1]

In Contextual Psychology, we believe that:

- Rather than resolving the past, we live fully in the present.

- There is no need to look back because the present context is really all we can resolve.

- Any presenting problem actually is a multiplicity—not only of problems, but of interpretations, experiences, behaviors, thoughts, feelings, actions, sensations.

- All of these things need to be addressed in order to us to find resolution and establish a goal or pathway for our clients.

---

1 Association of Contextual Behavioral Science http://contextualscience.org/contextual_psychology (accessed March 8, 2014 2014).

The context for Contextual Psychology is right now, this moment—the only thing that we can impact. In other words, Contextual Psychology differs from the approach of Neo-Freudianism in that regression (or regression-to-cause) in resolving the past is not viewed as the central approach to therapy. This is of course logical. In the present time, it's impossible for me to change the past. Yet the current approach that many therapists have to therapy is to spend time "processing the past".

The reality is that we can process the past forever. We will never be able to change the past. One of Freud's central ideas was that we could go back into the past to find the cause of today's issue. He thought we could resolve that past problem through a series of different processes and then experience life more fully in the present. It made for a great theory in the early 1900s, but the evidence shows us that we can't impact the past at all. In fact, the evidence shows us that memory isn't particularly accurate or even useful in many cases; as a result, regression to a specific cause and resolving that cause is an ineffective method as contrasted with those approaches of Contextual Psychology.

To me one of the most amazing elements of Contextual Psychology is that we figuratively change the past by changing the present. It's not because the past is altered or because it's even resolved or because somehow the past is different. Rather, through the processes of acceptance you will learn in this book, we can change either the importance of the past or our interpretation of the past. The result is that we are no longer enmeshed with the troubling past, instead enjoying this present moment to its maximum potential.

Although in the term contextual behavioral psychology we have the word behavioralism, it differs from classic behavioralism—that of B.F. Skinner and other—in that it doesn't try to explain why people act as they do. One of the primary goals of early behavioralism was really to figure out the why. What is the mechanism that causes this? It's my belief that life is actually a cause for many of the complexities that my clients experience and that we can work with clients session after session after session trying to find the mechanism or the cause for why they do something, and even if we do understand the why they are afraid to fly or why they are depressed, *why* still changes nothing.

The focus of Contextual Psychology is not why do I do this, but instead how do I live fully in the present moment. For example, Acceptance and Commitment Therapy (ACT Therapy) differs from other CBT approaches in that rather than trying to teach people to better control their thoughts, using what Albert Ellis called "thought stopping techniques," ACT therapy teaches them to avoid fusion and enmeshment with thoughts, feelings, and sensations.[2]

This is really a huge shift in psychology. In the era of pharmacology, psychiatry in particular gives people a false belief: *If you take this pill, you will become un-depressed. If you take that pill, you'll become un-anxious. If you take this pill, you will become un-psychotic.* The expectation in our instantaneous world is that when we go see a therapist, we will stop having those experiences, so most people present in counseling with the desire to "stop" something.

ACT therapy does not have as its goal the stopping of anything, but rather the acceptance of those thoughts, feelings or sensations; it uses the process of avoiding cognitive fusion and enmeshment with troubling thoughts, feelings or sensations rather than the goal of stopping them. This is the ultimate in paradoxical therapy: Through the paradox of acceptance, true freedom is ultimately found. The paradox here is that when something is accepted as being just what it is, it then has no power.

Depression, loneliness, hunger, fear, or even withdrawal become unimportant when accepted. When it becomes unimportant, it becomes just what it is. It is then something experienced rather than something that I hate or I fight or I'm restricted or obsessed with. I can find freedom from suffering through that acceptance. Depression is not a problem. I work with a lot of obese clients, and hunger, once accepted, is not

---

2 Steven C. Hayes, *Get out of Your Mind and into Your Life* (New York, NY: MJF Books, 2011).

problem. I work with a lot of addicts who learn that withdrawal is not a problem. Loneliness is not a problem. An emotion, an experience, is only a problem if my client makes it one.

What these things actually do is they let me know that I'm a human being, not a human doing. I've said before happiness would suck if life had no depression. Security would actually suck if we had no fear to put it into perspective. Difficult times and experiences are part of any valued path. Difficulty and pain for our client are not to be avoided if one wants a truly meaningful life. They're simply things to accept because in acceptance we give them no power to control. What our problems do then is become parts of the pathway to becoming a full human being, participating fully in life.

Acceptance seems like it's such a difficult thing for us to move our clients towards, yet there are specific strategies and techniques that we can use with our clients to help them move towards a process of acceptance. Acceptance, by the way, does not mean "I like something, I endorse it, I wish it would happen, or I want it to happen to other people." That's not what it means at all. It simply means, "I can live fully in the present despite where I've been in the past."

The reason we can move our clients to this point, even though some of their experiences are truly awful, is that when we teach them the context of the present, it gives them self-control—which is what most of our clients are seeking through their unhealthy manifestations of symptoms and distressing behaviors.

Contextual psychology differs from the primary approach of psychiatry in a number of different ways. Our goal is not through a magical pill to resolve the problem. It's actually to help our clients be fully human by living in their problem, which paradoxically brings a tremendous amount of freedom. Contextual psychology also differs from the primary approach of psychiatry and psychology in that diagnostic labels are far less useful to the therapeutic process than perhaps in any other field of behavioral health.

It was probably at least 25 years ago that the National Association of Social Workers (NASW) began developing an assessment tool that they called Persons In Environment (PIE). It was really a major project within social work for about 20 years. In fact for many years, the Persons In Environment manual was on the main page of the NASW national organization website.

And I've always been a big fan of PIE and the reason why is that PIE is an excellent alternative to the Diagnostic and Statistical Manual. Instead of pathologizing people with problems, it assesses people within their environments to help us as professionals find the solutions that would be most helpful to them. That approach is consistent with the principles and goals of Contextual Psychology.

What's the Role of the Therapist in Contextual Psychology?

Contextual Psychology is largely an educational endeavor. I see myself—and you are going to begin seeing yourself—fulfilling the role of a guide, a mentor, a teacher, or a coach to your clients.

Now I'm a big fan of the coaching model within counseling and social work for a very simple reason: It is far more powerful to do something *with* the client, to introduce something to the client, than to do something *to* them.

This is often not what the client expects. They expect that we will give them medication and make their problem go away, or we will use our mystical, magical, psychological processes that we learned in graduate school and make them un-depressed or un-anxious or un-scared. In couples counseling, they expect we will make them un-unfaithful, or we will make them un-uncommunicative.

I've never found a way to change my clients. What I found instead is that when I do something with my client, it changes the client—not because of my power, but because I've helped them to discover the power within themselves to make those changes. And filling the role of guide, mentor, teacher, or coach is really one of the central themes throughout many of the different approaches to Contextual Psychology like DBT, ACT therapy, Solution-Focused Brief Therapy or other approaches of Contextual Psychology.

One of the chief characteristics of Contextual Psychology, especially Positive Psychology, is that it really tries to look at what's right rather than what's wrong. It builds on and utilizes the client's strengths instead of fixing what's wrong. It's not to try to get the borderline over the border or make the anti-social pro-social.

As a marriage and family therapist, I've worked with some couples over the years who have so many problems, it would be impossible for me to fix what is wrong with their marriage. As an alternative, I try to help them discover what is right and to utilize that as a tool for changing the present. It's so much easier to focus on what is right than it is to fix what is wrong.

I can't change the past, but what I can do is give this assignment to every couple I work with: I tell Bob and Bertha that when they leave my office what they're going to do is stop at the store and they're going to buy a notebook, a spiral notebook like a high school you might take to class. One book, two people. And each day they're going to write down in that book one thing about their partner that they value. It could be one word, one sentence or a short paragraph. Anything more than that is probably too much. One book, two people.

I tell them to put it in their bathroom—or in their garage, in the kitchen, in the bedroom, wherever it is that each of them walks through or passes by each day. I have them write their last name on the front, for example, they can write, "the Smith Family Treasure Chest." They don't have to do this assignment together, but with one book for two people, they can see what their partner wrote.

I can't fix what's wrong with their partner, but most couples have been saving up what are called marital green stamps. They've been focusing on what's wrong with their partner waiting to cash that in with adultery or divorce or rage or whatever it is people cash those emotional and behavioral green stamps in for, rather than saving up treasures which they can cash in for something valuable to themselves and to their partner. And so rather than fixing what is wrong with the couples whom I've worked with, I help them take what's right and utilize that as the way to compensate for the deficits that are important. It's amazing how when they focus on what is right with their partners, what's wrong with their partners becomes so unimportant.

Over the years, I've worked with a lot of personality disordered individuals. In fact, I've done a lot of training and a lot of workshops on personality disorders, and it's interesting to me the feedback I get. People say, "I don't know how you can work with personality disordered folks. They're so difficult to work with." In fact, before the DSM-5, back in the days of the DSM-4, we had them on their own axis. That is not true in the DSM-5, which doesn't have a multi-axial system, but we had them coded differently than our other psychiatric clients. The reason why was the belief that personality really doesn't change and that if they're personality disordered client, they're always going to be a personality disordered client.

And so perhaps because my training was during an era when that belief was the dominant belief, I stopped trying to fix them. But what I did was I tried to tap into what was right with each of them so that they could use that strength or core aspect of their personality to compensate for the deficits that were present. For example, my paranoid clients were cautious. I wish more of my clients were cautious. Caution is actually a great attribute that can be a strength.

My schizotypal clients, although they were bizarre, strange or eccentric, often were able to express themselves in unique and interesting ways, very creative. I'm convinced that many of the great inventors of our time were schizotypal, and that creativity was a strength of that personality. Sure, they wore stripes with plaids, but they invented all kinds of amazing technology.

The schizoid individuals want their strength. They're autonomous.

They're able to function alone. The borderline personality causes so much grief, particularly in the inpatient treatment setting, but they're flexible, mercurial, and adaptable. Those are awesome personality traits. The anti-social: if you've ever led group therapy and nobody is talking, call on your anti-social. They'll kick start the group. You may have to do some clean up, but they have some great personality strengths in being able to communicate exactly what they need, want, or feel.

The dependent personality disordered personality individual is loyal, and loyalty is a tremendous problem-solving attribute. The narcissistic client is named for the story of Narcissus who was enamored of his own reflection. He's looking into the pond and of course eventually falls into the water and destroys his own reflection. There are lots of problems with the narcissistic, but I wish more of my clients had high levels of self-esteem. This is not an area where I need to work with these clients because they understand their own self worth and their own self value. That can actually be a strength in therapy.

The histrionic clients know how to express themselves, and they do wish express themselves. I wish more of my clients would communicate how they feel or what they want or what they need. The avoidant client is able to work alone, and that's really awesome. My obsessive compulsive clients are able to function within rules, structure, and order. All of these are great personality traits, especially for somebody like me who's worked with clients often in the context of criminal justice work.

Positive psychology looks at what's right rather than trying to fix what's wrong. Taking an approach in therapy where we only focus on the assets present, rather than the problems present, makes therapy so much more enjoyable not only for me as the therapist, but for my clients in my office as well. Let me share with you a couple of recent clients whom I've worked with and cases that I'm familiar with.

One of the clients I was in my office recently was a 400-pound, 56-year-old, obese attorney. He's tried every approach to weight loss possible, every diet under the sun. At 56 years and 400 pounds, if he doesn't lose half of his body weight in the next 18 months, he is going to die.

That's the end result of being so obese. My primary approach with him is based in Contextual Psychology. Mindfulness and the art of mindful eating, these are techniques that I'm going to talk about in the successive sessions. For him, the success has not only been numerical—his weight on the scale or the size of his clothing—but has also been in his ability to function in his marriage and in his job as a corporate attorney.

A nine-year-old suicidal girl whose father committed suicide a couple of years back has reported some of her own suicidal ideas. She's been seeing a psychiatrist for the past two years who's been medicating her perhaps under the belief that she somehow has a Prozac deficiency. She has been working with the same therapist for the past two years as well. I asked the mother what the results of two years of therapy had been and she said really not much.

So I said, "Well, what is she doing as homework between sessions so that she can practice what she's learning in therapy and apply it to her real emotions in the real world?" And the mother said, "She's not getting that in therapy." I said, "Then it's time to find a Contextual Psychology practitioner, a counselor or a social worker, who can teach the skills of emotional self-regulation and acceptance, and help her to define and develop a valued path, even at nine years of age." You find that in Contextual Psychology.

Another recent client of mine was a 40-year-old amputee. He came to me for smoking cessation. I see a lot of people for smoking cessation. His most recent amputation had been a few weeks beforehand. It was still healing. He came to me for smoking cessation because it was not healing correctly. For him, smoking cessation was a manner of life or death. But when he came in, he was not only suffering the difficulties of his recent amputation, he also very clearly was heavily medicated by his physician.

And I thought to myself, "I'm not even sure that I can work with this person in my office because of his cognitive state." I spent probably about thirty or thirty-five minutes interviewing him, and at that point I was really ready to make a referral, not sure that I was going to be able to really help him with the type of strategies that I provide. Then I looked at his intake form and saw that the only hobby he listed was Bible stories.

I thought to myself that that was a rather odd thing to write down, so I said to him, "What type of Bible stories do you like?" He answered, "Oh really, I just sit around all day long being disabled, but these missionaries come by, and they leave me literature with Bible stories." It was real clear that my religious faith and his religious faith were two entirely different religions. But my goal in therapy is not to make my client the best me that I can, but I help them become the best them that they can be.

And so I asked him a question. I said, "What stories are most interesting or important to you?" And he told me a couple of stories, and because I have degrees in ministry in addition to therapy, I knew the historical roots of these stories and so I spent the next two sessions with him using a metaphorical and story-based process which we're going to examine extensively in this book on Contextual Psychology. He ended up not only not smoking, but healing quickly from that amputation. He was able to move towards a state of acceptance related to the next medical procedure that he was going to have to endure. He left my office not only physically better but emotionally better. His wife was extremely grateful for the time that he had spent in my office.

I recently worked with a 32-year-old medical patient who was afraid of needles and medical procedures. A recent diagnosis had resulted in her having each week to endure certain medical procedures that involve phlebotomy and needles, and this was causing her a lot of anxiety and panic. In one session I was able to help her find a significant level of relief that allowed her to undergo the process the next week. In my second session with her, I was able to ratify the change that she had made. In simply two sessions, she experienced the success that she needed to continue on with her medical treatment.

As you prepare to learn the ins and outs of Contextual Psychology, you can begin to think about how you will use these ideas and techniques with the clients on your caseload.

CHAPTER TWO

Relational Frame Theory

When I was telling another colleague about this book, they imagined that Relational Frame Theory would be the last thing I would address because it's a little on the obscure side. But Relational Frame Theory is really the starting point for understanding the methods of contextual behavioral therapy.

No matter what approach you use, the relational frames that your clients create are really the heart of our focus in change work. Relational frames are the reason that in couples counseling, two marriage partners have entirely different interpretations of the exact same experience. Relational frames produce the automatic behavior that we see in our clients related to unhealthy coping. This ranges from drinking to cutting and self-mutilation to isolation from others and really just about every other automatic, unhealthy self-defeating behavior.

Relational frames are the cognitive evolutionary trait that puts mankind at the top of the food chain. They are what really separate us from all of the other mammals and all of the other species on the planet earth.

Relational frames at the same time are perhaps our greatest cognitive deficit. These are the things that keep people from reaching their greatest level of potential. It is the relational frames that our clients create that bring them to our office as psychotherapists. So that of course brings up a really important question. What are relational frames?

Relational frames are the mental and often subconscious/unconscious constructs that support an idea, a belief, an experience, an interpretation, or an action.

When you think of a tent, you probably think of the canvas. You're sort of picturing it in your mind, looking at the canvas of a tent. Maybe it's orange canvas or maybe it's green canvas or some other color of canvas tent that you think of as a tent, but it's actually the aluminum tubes that are the frames supporting it, holding it up, and making it the experience of a tent. Without that frame, without those aluminum tubes supporting that canvas that we think of when we think of a tent, it's really just a pile of canvas.

Without mental constructs, mental frames, then ideas are just ideas. Experiences are just what is, and feelings are just feelings. Have you ever thought about what makes anger feel angry? It's really a great question. It is the frame that we've created for understanding anger that makes anger angry. Anger means different things to different people because of the various relational frames that they put that anger on.

For our different clients, anger can mean any of these things:

- I'm sad
- I'm hurt
- I must cut on myself
- I'm bad
- I'm worthless
- I am right

Have you ever bought one of those watches with the different colored interchangeable bezels or the different bands? I have a 6-year-old girl at home, and I bought her an interchangeable watch band bezel set. It's very colorful, and she has enjoyed playing with it.

It's the same watch though no matter what band or what bezel is on it. But when the frame changes, it changes the watch. This is really an analogy or a simile that can help us to understand relational frames. Relational frames can be kind of tough to get our heads around because relational frames are often subconscious, arbitrary, and something that we've learned. And when we talk about learning, we're talking about something that we've learned really on a lifelong basis.

In fact in many cases as a family therapist I can speak to my clients learning these relational frames on an intergenerational basis. Anger meant the same thing to grandpa that it meant to daddy that it means to me. So, when I tell my client that that is not what anger means or how anger should be expressed, they look at me like I'm from Mars. I spent the first ten years of my career in counseling working with drug addicts and alcoholics, and when I told them the way they drank or the way they use drugs was not normal, they said, "What do you mean? This is the way everybody drinks" or "this is the way everybody uses drugs." I had to explain to them how it's actually only the way alcoholics and drug addicts drink or use drugs. It was something that was a learned pattern, and they believed because they learned these relational frames that that's what normal was.

Let me come back to the word arbitrary because I think that's really pretty important.

In fact, arbitrariness is one of the most unique features of the human mind. It's what actually lets us be creative. It's what gives us ingenuity and has evolved mankind from the Stone Age to the Computer Age. Arbitrariness is what corporate trainers seek. They're always looking for somebody in corporate training who can "think outside of the box." The problem is that this talent also seems to happen indiscriminately.

We are always mentally scanning the warehouse of the subconscious mind to find the right frame to hang our present experience. This is often the context in contextual therapy. We do this without thinking about thinking. Probably the best example of this is a fish. A fish does not know that it is swimming in water. It has always been in water. A fish does not think about water or know what being wet is, because a fish exists only in water.

Similarly, we're swimming in our thoughts. It's what we always do. These thoughts we are swimming in are searching every second of our waking life for the frames to make sense of our experiences and what's going on around us.[3]

So what does this have to do with counseling or therapy? It sounds kind of interesting from a theoretical perspective, but our primary task in Contextual Psychology, our primary task in counseling and social work, is really to help people un-frame their thoughts, un-frame their feelings and un-frame their sensations. Why?

Well, because they're often wrong. The frames have been arbitrarily created and applied to the experiences, and this creates cognitive errors. We've all learned in the past that the basic task of Cognitive Behavioral Therapy is to counter cognitive errors, but how can this be done if the client still holds on to the arbitrary frame that they've hung the experience on? Even if we get confrontational Gestalt therapy and confront the present cognitive error—think of old time substance abuse counseling where we had a round circular group and put somebody on the hot seat so we could confront their denial or their cognitive errors—our client is likely to hang the next experience on that same frame. This is where relapse comes from.

As a therapist, have you seen these types of relapse?

* relapse into drugs
* relapse into depression
* relapse into panic after being on medication for a long period of time

---

3 Ibid.

- relapse into panic after being in therapy for a long time
- relapse into obesity
- relapse into old patterns of communication and fighting (in couples therapy)

If you have seen this, learning how to deconstruct relational frames is really as important as any other task in the counseling process.

Some readers will know that in addition to being a family therapist, I'm also a certified clinical hypnotherapist. I always tell people that my job is un-hypnotizing people as much as it is hypnotizing them. As a matter of fact, I actually spend the majority of my time as a hypnotherapist doing un-hypnosis with people breaking these relational frames. As a therapist, my job is to help people let go of their thoughts as much as it is to help them create new thoughts. In community counseling or a social work setting, our job is often to help people and organizations to find new frames--those compatible with the organization and community goals. Our job is often to help an organization or a client recognize that just because a frame exists, it doesn't have to be used.

Understanding Relational Frame Theory in the context of cognitive behavioral therapies leads to a therapeutic approach then that's really very flexible because it moves with your clients' thoughts. As a therapeutic approach, it is great because it fosters curiosity: not only curiosity on my part, but my clients learn to be curious about themselves.

RFT leads to a therapeutic approach that is mindful, one that is focused on the present because when we slow down enough to truly experience this moment, then we don't have to quickly and arbitrarily create frames that may or may not be useful. Relational Frame Theory leads to a therapeutic approach that is solution-focused. It's focused on a specific outcome that's beneficial to our client, change work that actually has meaning.

Understanding Relational Frame Theory leads to a therapeutic approach that teaches our clients skills that can be replicated. My goal in therapy is not to help my client come back next week for another session. Ultimately, my goal in therapy is to get rid of my clients. I've spent a lot of time learning how to do effective marketing, and I've shared those techniques in business development classes for therapists. Every now and then I'll meet somebody who will say to me, "Why the emphasis on marketing?" and I say, "Because that's how I know I'm a good therapist. I know I'm a good therapist because my clients don't come back to me."

The reason I use that as a benchmark for success is because my goal in therapy is to teach my clients something that they can do on their own apart from the therapeutic process, which is why all of my clients get homework assignments.

Relational Frame Theory as a therapeutic approach really helps me to be empathetic. When I understand Relational Frame Theory, then I can see people as they are, rather than as they should be or could be or ought to be. It removes judgment from the process, helping me to recognize that even those who find themselves in extremely difficult circumstances—in many cases because of their own choices—are people who deserve to be worked with. There's hope for helping those people make tremendous amounts of change.

While writing this chapter, I encountered an interesting example of a relational frame. Stephanie, who is my company's administrative and technical support genius, called and asked me for some documentation—documentation that she already had.

I knew that she had the information because it's in a folder that she uses on a daily basis with the instructions for all of the other courses that we offer. When I directed her to that folder, she laughed and said, "You'll have to be gentle with me today. I just put my son on a plane." Her son would be spending the next six months overseas in a training program for a missionary group called Youth with a Mission. Now that's exciting, but as any parent knows, sending your nineteen-year-old overseas for six months to study can be pretty difficult.

The relational frame that Stephanie created was completely arbitrary. She came to the conclusion rather rapidly and quickly in her own mind that the reason she was the absent-minded professor that day was because she was dealing with emotions related to empty-nest syndrome.

Experiencing Relational Frames

Do you want to test your mind's power to create relational frames? Relational Frame Theory can be pretty complex from the psychological perspective, so I often use an exercise as an easy way to wrap our head around Relational Frame Theory. It comes from one of my favorite client resources, *Get Out of Your Mind and Into Your Life* by Steven Hays. This exercise will show your own ability to create these arbitrary relational frames. You'll need a piece of paper and something write with.

1. Pick out any two objects that you can see right now.
2. Write them down on the paper.
3. Write down your answer to this question: How is the first object like the other object?
4. Write down your answer to this question: How is the first object better than the second object?
5. Write down your answer to this question: How is the first object the parent of the second one?

For this exercise, I picked two objects: the first is headphones, the second one is car.

And so question number one: How is the first object like the other object? For me, I thought about it for a moment and I realize that you can listen to music with both of them. This is an example of our mind's ability to really come up with an answer for almost any question.

Now, question number two: How is the first object better than the second object? I came up with the answer that you don't have to buy gas for headphones. It's not expensive. Again, that relationship is completely arbitrary.

Finally, question number three: How is the first object the parent of the second object? The answer I came up with might be bizarre, but it is this: You listen to headphones sitting in the chair that the father sat in on the TV show *Leave It to Beaver.*

It may be a bizarre answer, but it's the answer that's satisfactory to me. The first object is the parent of the second object because it is used in the chair that father sat in the TV show *Leave It to Beaver.*

No matter how bizarre your answer is you can always come up with an answer. Your abstract thinking skills actually create an answer that at least at some level will make sense within your mind.

Now, let's apply some logic to this. Can everything be the parent of everything else? Of course not. It's not possible—and yet in our own minds, we have the ability to create these relationships even when they aren't logical, even when they don't make any sense.

This is what happens in the therapy room. This is where cognitive errors come from. This is the heart of problem-solving in many approaches to contextual therapy. It may be an awesome ability to create these relational frames, but it can also be a huge liability for the clients on our caseload.

Delving Further into RFT

Let me at this point turn this into rocket science and explain Relational Frame Theory a little bit further.

The main proponent of Relational Frame Theory is Steven Hayes, a psychologist from the University of Nevada who's published an incredible amount of peer reviewed literature on the efficacy of ACT therapy. He considers RFT a foundation for understanding how to do Contextual Psychology.

His purpose in exploring relational frames was to go beyond the work of B.F. Skinner, who never could really explain how language fit into the equation. Humans express very complicated ideas. They can often do this with very little communication; the mystery as to how that can happen was always really the thorn in the side of B.F. Skinner and classic behavioralism.

To explore RFT a little bit further, we need to understand some of the principles of language. The meaning of a word is truly arbitrary. It is assigned by collective agreement. In my home, there are four different languages spoken—maybe five if you want to count my wife's tribal language—so I have a lot of experience listening to people who use different words to mean the same thing.

Mom means mom because that is what the culture decided would mean mom and really for no other reason. The meaning of words truly is arbitrary. The meaning of words is not inherent. Meanings are actually inferred, and because of relational frames we can identify relationships between words and meaning, so *anger* means something or *depression* means something else through inference.

Just like the meaning of words, the answers to the three questions in the earlier exercise come entirely from imagination and inference. They might make sense only to the individual playing the game, and only in that moment, and yet our minds reach those sudden inferences almost effortlessly.

According to B.F. Skinner's operant conditioning, we learn only from experience, from repetition. RFT shows us that we also learn from the creative nature of the subconscious mind. Those familiar with Cognitive Behavioral Therapy may recognize this as the foundation of self talk.

Perhaps the easiest way to explain Relational Frame Therapy is by thinking about a child and a dog, and the differences in their language, ideas, and abstract abilities. If you give a dog a biscuit and you say *biscuit* when he's munching on it, the dog now associates the sound biscuit with the treat. All you have to do now is say *biscuit,* and the dog will come running even from the other room.

Now, you and I know that a biscuit is also a treat. We understand this because we're human, and we use language. So now suppose that you stand in one room and yell out *treat!* Unless you've previously associated the sound *treat* like we did the word *biscuit* through repetition, the dog is probably going to ignore you and stay in the other room, continuing to look for the cat. After all, the cat is far more interesting than a sound the dog has never heard before.

Now, give a child a biscuit. I suggest giving him one of those awesome Biscotti biscuits. Those are my favorite. That's a double big cookie. So, if you give a child a biscuit and you say *biscuit,* the child will now associate the sound *biscuit* with the treat. Now, go into the other room and yell *biscuit,* and the child will come running, even from the other room just like the dog did.

Later on, you can yell out *treat,* or *dessert,* or *sweets,* and the child will come running.

The dog didn't do that, but the child did. Why? Because the child arbitrarily creates a relational frame so that many other words that we might use will be arbitrarily associated with the delicious taste of the Biscotti biscuit.

So think about some of the words we hear in couples counseling:

- adultery
- sex
- hurt
- fear
- cheater
- fired

- support

- alone

What the words mean to our clients are a function of their relational frames. These words are meaningless to a dog, which is why perhaps we don't have dog therapists, but they're meaningful in couples counseling and have different meanings to each partner that we're working with.

Even though RFT is a theory, not a therapeutic method, understanding Relational Frame Theory enables us to understand the cognitive processes that trip up our clients. RFT underpins the treatment plans we use in couples, individual, group or even organizational counseling. My hope is that by exploring the concept of Relational Frame Theory, you can learn how important words are in the therapeutic process., as well as how important meaning is.

Contextual Therapy can be understood as a process of creating interventions will help our clients break non-resourceful relational frames and avoid the creation of new unhealthy relational frames. As a therapist, my job is to un-hypnotize people. It's to move them from arbitrary subconscious associations really to the power of conscious living. And the rest of this book will be devoted to the techniques that can help us do that.

CHAPTER THREE

Challenging the Idea of Past-Tense Therapy

Before we go any further in our study of Contextual Psychology, let's challenge a predominant viewpoint in therapy. In the hypnosis community, it is called regress-to-cause, but it has other names and crops up in many different forms throughout the various therapeutic profession.

Sigmund Freud originated the basic idea of going back to an emotional root cause and creating change by provoking catharsis. At its essence, this is past tense therapy.

Now, I recognize that most of us don't identify as Freudians, but his influence was pervasive across many years in our profession and continues to be. To a large extent, the ideas of Sigmund Freud are still a part of our everyday vocabulary in the counseling office, not to mention a part of the frame our clients often bring to therapy. For example, do you use these questions or phrases with your clients:

- So tell me what happened?
- Let's go back and take a look at that, what it felt like.
- How did you feel when… ?

- Go back now maybe a week or two, maybe a couple of months, maybe a couple of years, maybe to a specific event or feeling or emotion and let's talk about when that happened.

- And what did you do when _____?

These are past tense approaches to therapy. These are questions that therapists ask because at some level there remains the belief that in order for me to be okay I have to go back into the past and either resolve or relive or have catharsis or re-experience or reinterpret something that already happened. This whole philosophy is based on the ideas of Sigmund Freud: that somehow our past controls our present. Freud believed that neurosis is caused by unresolved conflict in the past and that revisiting the past will change the present.

The primary mode of operation that Freud used then was catharsis related to those events. Josef Breuer was a contemporary of Sigmund Freud, an Austrian physician as well. The Breuer and Freud theory was that symptoms are caused by repressed emotions, not repressed as in forgotten, but underlying in the subconscious mind.[4]

Freud writes in *Studies in Hysteria* "Each individual hysterical symptom immediately and permanently disappeared when we had success in bringing clearly to light the memory of the event by which it was provoked and thus arousing its accompanying effect."[5]

In his later work, Freud actually looked back on these early cases and he really wasn't completely satisfied with the results of catharsis. In fact, by the end of his career he actually was dissatisfied with the whole cathartic component of therapy.[6]

Freud's major contribution from a theoretical perspective—the ego, the id, and the super ego—was actually made in his latest years of life,

---

4 "Internet Encyclopedia of Philosophy", University of Tennessee http://www.iep.utm.edu/freud/ (accessed Mardh 8, 2014 2014).
5 Sigmund Freud, Josef Breuer, and Nicola Luckhurst, *Studies in Hysteria* (London ; New York: Penguin Books, 2004).
6 Ibid.

but it seems as if his early theories of repression and catharsis are still the predominant legacy of Sigmund Freud.

In the early 1970s, Arthur Janov elaborated on Freud's ideas.[7] He claimed that if infants and children are not able to process painful experiences fully—for example to cry, sob, wail, and/or scream in a supported environment—then their consciousness splits.

Pain gets repressed to the unconscious and reappears later in neurotic symptoms and disorders later in life. According to Janov, painful experiences become stored. They need to be released in therapy by relieving and discharging suppressed feelings. Janov claims that cathartic emotional processing of painful early life experiences and the process of connecting them with the memory of the original event could fully free clients from neurotic symptoms.

But Janov and certainly his predecessor Freud and Breuer as well as a number of others were really theorists. These were not empirically validated techniques, at least not by today's contemporary peer reviewed standards. In fact, the vast majority of Sigmund Freud's writings are simply case studies, observations that he made.

In contrast, when we're talking about contextual behavioral psychology, we're talking about evidenced-based treatment. We're talking about literally thousands of peer review journal articles measuring the efficacy of the therapeutic approaches. While none of us would be here without people like Janov, Freud, and Breuer, the reality is these approaches are now regarded as defective approaches.

Here are some of the problems with these past-tense approaches whose legacy still lives:

* Regression: Regression is only a metaphor. It's not actually a reality. It's impossible to see things today from the same vantage point as yesterday. The whole idea that we can go back, see

---

7 Arthur Janov, *The Primal Scream; Primal Therapy: The Cure for Neurosis* (New York,: Putnam, 1970).

something, and experience it as we experienced it then is really just a metaphor. It's not an actuality.8

- Memory: Memory itself in unreliable. It's flexible and adaptable. Elizabeth Loftus from the University of Washington has established in numerous studies that what the mind cannot recall, the mind will create. This is why you can have two siblings who have an entirely different recollection of memory of what their childhoods were like. Memory is notoriously unreliable, not only for adults reviewing what happened when they were children, but also in the couples who are reviewing what happened last week.

- Narcissism: It's rather arrogant to suppose that the awesome therapist will be able to review a person's life and their history to pinpoint "the cause" of today's distress. It's really a rather narcissistic presumption in therapy. In fact, it's my belief that life is the cause of the problems. For all of my clients presenting problems, there's usually not a cause. There are usually a multitude of causes. Contextual psychology tells us that cause is not something we really need to be concerned about in order to experience change.

- Lack of Effectiveness: The research shows the methods of psychoanalysis and free association to be only moderately effective. They have been discarded in large part by psychiatry, psychology, and medicine for the better part of the last fifty years. To continue to embrace what I call past-tense therapy approaches really denies logic, history, and the overwhelming evidence of today's research.

- Labor-Intensity: Focusing on the past is a labor-intensive endeavor. It's not consistent with short term therapy. We live in the world of managed care. Sigmund Freud's ideas was if the client comes and sees a therapist twice a week, every single

8 T. X. Barber, "Hypnotic Age Regression: A Critical Review," *Psychosom Med* 24, (1962).

week, for a period of ten years, at the end of the ten years, the client and therapist will have restructured the client's personality. Supposing for the sake of argument, that was a possibility, the reality is in the era of managed care, limited time and limited resources, none of us have the time to focus on the past.

So, really what is the alternative?

Present-Tense Therapy is the alternative—or as it's better known, Mindfulness.

CHAPTER FOUR

Mindfulness as a Success Foundation

Contextual Psychology is the antithesis of Sigmund Freud's approach. Contextual approaches are present-focused. They don't search for a cause, but rather they train the client in a new skill, the skill of acceptance.[9] While many traditional approaches to therapy have dwelled on the past, the foundation for Contextual Psychology is helping our clients to live fully in the present.

Mindfulness is an important skill taught to clients. With the vast majority of my clients, I teach them a basic strategy for Mindfulness within the first thirty minutes of the first session.

Interestingly, the importance of Mindfulness is not usually obvious to my clients during the first session. None of them say, "Wow, that was incredible, Richard. That was the most awesome experience I've ever had." In fact, when I'm done in that first session teaching them a basic strategy for Mindfulness, they almost always respond with, "Well, I guess that was relaxing," or "Uh-huh, I guess I can see how that could be helpful."

---

9   Steven C. Hayes, Kirk Strosahl, and Kelly G. Wilson, *Acceptance and Commitment Therapy: The Process and Practice of Mindful Change*, 2nd ed. (New York: Guilford Press, 2012).

The value in Mindfulness is not in guiding a client through a basic process in our office. The value of Mindfulness is in teaching them a skill that they will practice on their own between sessions so that they cultivate living fully in the present.

In the second, third, or fourth session when they come back, there comes a point when they say, "Because I've been practicing everyday, I now recognize why the first session was so important. I'm actually living mindfully. In situations where I used to automatically become anxious, I am automatically being mindful. In situations where I would respond with an impulsive behavior, I'm responding with mindful awareness."

So it's important to keep in mind that Mindfulness exercises are not the point. Learning to live mindfully is the point.

Mindfulness is not a new idea. In fact, the basic tenets of mindfulness are probably 5,000 years old. However, the application in psychology is new, because psychology is new. Up until a hundred years ago, we always understood human behavior in context of religion and theology. Buddha did not know that he was starting a religion called Buddhism. Buddha really was to a large extent one of our first psychologists. (Indeed, many Buddhists insist it is not a religion, but a philosophy.) Buddhism, being one of the first attempts to understand the human mind, recognized the value of cultivating Mindfulness, and so sometimes when people hear about Mindfulness, they think I'm going to be teaching Buddhism.

Usually by the way, it is not my clients who wonder this, but usually it's other therapists who wonder if I'm a Buddhist psychotherapist. I'm not a Buddhist psychotherapist. In fact, if I were to teach Mindfulness to those who are Buddhist, they would say that's not Buddhism. Teaching an idea, a technique, or a strategy for problem resolution that originated within a theological frame of reference is not the same as teaching that religion.

Simply put, we've been trying to understand ourselves for the last 5,000 years and most people throughout history have tried to understand that through a theological lens and so it's no great mystery to me why Buddhism addressed a subject that really is ultimately about psychological health. By the way, Buddhism certainly isn't the only religion to address the concept of Mindfulness: It is present in the work of Jewish meditation, Kabbalah, Islamic meditation, Christian meditation, Sufism, the ancient Roman Stoics, and many other schools of wisdom.

In fact, when Jesus talks about the lilies of the field not needing to worry about what clothes they will wear (Matthew 6:28), it's really the same concept of Mindfulness that other religions have talked about. So, Mindfulness is not in and of itself religious, even though religions from around the world have certainly employed the concept of Mindfulness.

Here are some definitions of Mindfulness:

◆ Bringing one's complete attention to the present experience on a moment-to-moment basis.10

◆ Mindfulness is paying attention in a particular way on purpose to the present moment and nonjudgmentally.11

◆ The first component of Mindfulness involves the self-regulation of attention so that it's maintained on immediate experience, thereby allowing for increased recognition of mental events in the present moment. The second component involves adopting a particular orientation towards one's experience in the present moment.12

◆ An orientation characterized by curiosity, openness, and

10 William R. Miller, *Integrating Spirituality into Treatment : Resources for Practitioners*, 1st ed. (Washington, DC: American Psychological Association, 1999).
11 Jon Kabat-Zinn, *Wherever You Go, There You Are : Mindfulness Meditation in Everyday Life*, 1st ed. (New York: Hyperion, 1994).
12 Scott Bishop et al., "Mindfulness: A Proposed Operational Definition," *Clinical Psychology: Science and Practice* 11, no. 3 (2004).

acceptance.

The last one is really my favorite definition of Mindfulness because I think it describes fully what my clients need to know in order to experience change in therapy.

The idea of Mindfulness and staying in the moment might sound simple, but these are skills that need to be taught to a client. It is not something that happens organically just because they come to therapy each week. A big part of the homework that I assign to my clients is to practice two minutes of Mindfulness with intention each and every day between now and their next session. By the way, that's that great thing about Mindfulness. It's not about meditating for thirty minutes while assuming funny postures or wearing funny clothes. In fact, a person doesn't even need to be still in order to practice Mindfulness.

You can mindfully shop at the grocery store. You can mindfully walk. You can mindfully be still in a chair for a period of time, and you can mindfully eat raisin. Mindfulness is not about a specific style or vantage point of meditation, and it certainly is not about clearing one's mind. The goal in Mindfulness is not to empty one's mind or to stop thinking. In fact to the contrary, it is to simply give us time to allow ourselves to observe ourselves swimming in our thoughts.

A Basic Mindfulness Meditation

While you could read volumes about Mindfulness, the best way to learn about it is to experience it. I will provide a script similar to the process I use with my clients; to truly learn about Mindfulness, you should with intention practice Mindfulness each and every day, twice a day for the next seven days. That's the same assignment that I give my clients, and I guarantee that if you do the assignment then you will find what my clients find: when they cultivate Mindfulness, the value of it is incredible.

This is the Mindfulness meditation week one practice. Like an athlete or musician, practice is for a performance. Mindfulness meditation is our daily practice for living life to its greatest potential.

There are three components to the practice:

1. *First, the practice of directing your attention to your breath.*

2. *Second, practicing how to return your attention to your breath anytime you notice feelings, thoughts or sensations. The goal is not to stop thinking, stop feeling or to stop having sensations. The purpose is to simply note when you do this and to practice bringing your attention back to a focal point, in this case the breath.*

3. *The third part of this practice is to begin to notice how easy and natural it is to stay in the present when we notice our attention drifting into either the past or the future. Notice during this week times when you mindfully and intuitively return from distressing thoughts, feelings, or sensations back to your breath into the present. As you sit in your chair with your body relaxed and your posture in alignment, close the eyes and breathe in noticing what it feels like to breathe in.*

Scan your body and loosen any muscles that are holding tension. Relax the jaw and let the shoulders drop and you can let your eyelids and hands be heavy with relaxation as you just breathe. You don't have to try to speed up or slow down the breath during this exercise. All you have to do is breathe and pay attention to your breath. There's not really a right way or a wrong way to do this exercise. It's simply the practice of bringing your attention to your breath.

Observe the breath, noticing the tempo of your breath, the temperature of the air. Observe how the air flows in, and what it feels like to flow out. We breathe everyday often without noticing it and by practicing an awareness of the breath, we're really practicing an awareness of this moment. After all. this moment, this breath is really all we have, and of course as long as we're breathing in this moment, we're okay.

As you breathe in and out, label the breath. Call it by its name. Label the in-breath "in" and call the out-breath "out." Say to yourself "in" and "out." Notice the air as you breathe in and the point where the air in your lungs turns around and becomes an exhale. As you pay attention to the breath, you'll also notice you are aware of sounds, sensations, and experiences apart from your breath. The practice is not to stop noticing those things, but rather when you notice thoughts and awarenesses, outside of the task of paying attention to the breath, simply note that you're doing that and return your attention to the breath.

If you notice yourself thinking about anything at all, you don't have to try to stop thinking; rather just note the thought instead of following it. Simply say to yourself that is a thought. Bring your attention back to the breath. If you become aware of an emotion or a feeling during this time, it's okay to have them. The practice here is not to suppress them, but not to follow them, to simply note them and say "that is a feeling" and return your attention to the breath.

Likewise, if you have any sensations, if your body feels something, you can simply note that is what my body feels, that is a sensation and experience and without becoming engaged in it or following it just use it as a cue to return your attention to the breath noticing what it feels like to breathe in and out. Over the next two minutes, continue to breathe in and out paying attention to your breathe.

The practice is of course to simply note when your mind begins to follow a thought or a feeling or an awareness of sensation and

*to gently, without judgment, return your attention to the breath.
It doesn't matter if you need to do this many times. The value
is in the developing the practice of returning to this moment by
returning your awareness to this breath and begin now. Spend
about two minutes doing this, and then pay attention to the next
breath, reorienting yourself to the floor below you, the air in the
room around you and opening the eyes.*

Note: I have an audio version of this posted on YouTube, which
you can access and even share with clients by visiting: http://youtu.be/
VEDPsFznX3s

Although almost everyone can see the value in this exercise, most
will discover the real value comes with practice. During this week as
you practice this exercise, begin to be aware of and notice when you
intuitively practice Mindfulness and mindful awareness of the moment
and other situations during the week. For example, if you're stressed in
traffic, you might notice when you become aware of the stress that you
can automatically focus on your breathe rather than letting the stress
become a thought you follow.

By the end of your first week of practice, you'll begin to notice how
natural and easy it is in a variety of different situations and places to
mindfully focus on one minute at a time, one moment at a time.

Mindfulness as an Evidence-Based Practice

What does a research tell us about Mindfulness? You'll find for almost
every area where we work as clinicians, there is a mountain of evidence
showing the efficacy of Mindfulness-based therapies in helping people
to experience success.[13] Rather than review the most current research at

---

13 Ruth A. Baer, *Mindfulness-Based Treatment Approaches : Clinician's Guide to Evidence Base
and Applications*, Practical Resources for the Mental Health Professional (Amsterdam ; Bos-
ton: Elsevier, Academic Press, 2006).

the time of writing this book, I'm going to tell you how to find the most current research when you read and re-read this book.

Go to www.scholar.google.com, which gives you only peer-reviewed journal articles, textbooks, and other academic publications. This is a great resource for any therapist. You can search the current literature. You can search the history of literature.

I don't know what type of clients you work with, but just type in the type of client you work with and the word *Mindfulness*. For example, I work with smokers so I might type in "smoking cessation Mindfulness" in the www.scholar.google.com. If I type it in the regular www.google. com, I'll get a bunch of people selling meditation CDs, but if I go to www.scholar.google.com, I'll find out what the research actually says about using Mindfulness as a strategy for helping people stop smoking.

Here's some of the really universal outcomes of teaching Mindfulness to clients. It seems like a simple strategy, but I've learned that sometimes what's simple is really most effective for my clients. Clients who are taught and practice Mindfulness reduce their rumination. Now, again, because I'm a marriage and family therapist, I'm really familiar with the problems of rumination. With most of the couples I work with, their problems are directly tied to one or both of them ruminating about the other person for the past five to fifteen years.

My depressed clients are ruminating about their depression. My anxious clients are ruminating about their fears. Mindfulness reduces rumination, and the research shows that. It reduces stress. That's really important for my medical clients.

I see a fair number of medical clients with significant complications. In fact, the book I wrote on medical meditation quoted 297 different studies that demonstrated that those who learn Mindfulness prior to surgery have three predictable outcomes[14]:

---

14 Richard Nongard, *Medical Meditation: How to Reduce Pain, Decrease Complications and Recover Faster from Surgery, Disease and Illness* (Scottsdale, AZ: Peachtree Professional Education, Inc., 2011).

- One, it decreases complications.
- Two, it increases the speed of recovery.
- Three, it decreases the dependency on medications.

Think about the diabetic clients you work with. Think about the clients with chronic pain. Think about the clients who work in a stressful job situation. For those of you doing therapy in EAPs, Mindfulness is going to be one of the most effective strategies you can teach clients. Mindfulness actually improves memory function. It enhances focus.

I don't work with adolescents anymore, but I did for a number of years. The only adolescents whom I do see are top-performing students who need help with test-taking anxiety. I use Mindfulness to help them enhance their focus.

Mindfulness, the research shows, decreases emotional reactivity. Think about the clients whom you work with. Do you wish that they could decrease their emotional reactions to the experiences they have? Many of our clients are in our offices because they don't have the emotional control over their reactions, yet the research shows those who learn Mindfulness have decreased emotional reactivity. They have greater cognitive flexibility. They're able to consider new and different options. Those documented researches show that those who learn Mindfulness improve their relationships. This can be and should be a chief strategy in marriage and family therapy.

There are lots of journal articles on applications of Mindfulness to pain control, eating disorders, ADD, ADHD, impulsiveness, marriage counseling, anxiety, personality disorders, depression and many other commonly treated issues in therapy.

When therapists practice Mindfulness with intention, do they derive a benefit in their therapeutic practice? There's actually research to show it increases a therapist's self efficacy,[15] their ability to develop attentional processes to increase their own patience and intentionality with clients. It helps them develop skills, which actually makes them more effective.

Therapists who practice Mindfulness become excited to pass along this strategy to others while developing skills that make them more effective as therapists: increased empathy and the development of a nonjudgmental compassion. For that reason, I'd like to encourage you to practice the Mindfulness exercise in this chapter twice a day for the next seven days.

Further Reading on Mindfulness:

Full Catastrophe Living: Using the Wisdom of Your Body and Mind to Face Stress, Pain and Illness by Jon Kabat-Zinn. Jon Kabat-Zinn is a neuropsychologist at the University of Massachusetts Medical Center who's been working in the area of chronic pain for the past 30 years.

Here You Are: Discovering the Magic of the Present Moment by Thich Nhat Hanh, a Nobel Peace Prize nominee and Buddhist monk who lives in France. One of the things that I really appreciate about Thich Nhat Hanh is that he explains the ideas of Mindfulness in a very non-religious way.

Mindfulness in Plain English by Bhante Gunaratana is another textbook which is often recommended by people. It is a book that I have found particularly useful as well.

---

15   D. M. Davis and J. A. Hayes, "What Are the Benefits of Mindfulness? A Practice Review of Psychotherapy-Related Research," *Psychotherapy (Chic)* 48, no. 2 (2011).

CHAPTER FIVE

ACT Therapy Solutions

[Editor's Note: Steven Hayes has stated that even though ACT Therapy stands for Acceptance and Commitment Therapy, it is an acronym pronounced "ACT Therapy" like "actor" as opposed to the initialism "A-C-T. Therapy," pronouncing each letter.]

Acceptance and Commitment Therapy (ACT) has been developed in large part by Dr. Steven Hayes, a researcher at the University of Nevada. Extensive studies verifying its efficacy have established it as an evidenced-based treatment protocol.[16]

There really are no gurus in Contextual Psychology. Certainly there are researchers whose research we might quote, but that doesn't make them gurus. In fact, Steven Hayes in Acceptance and Commitment Therapy is the perfect example of a transformational leader in our profession. Why? Is it because he's gotten everything right? Absolutely not. At various points, his work has certainly faced a wide variety of challenges from an empirical perspective.

Steven Hayes is a great exemplar for transformational

leadership because he hasn't said, "This is about me." Instead he has trained a cadre of leaders are experts in taking ACT Therapy to the next level. It's always amazing to me whenever I see Steven Hayes has published a new book. It is almost always a collaborative effort with other researchers in the field. There are usually three to six co-authors with him in almost all of his materials. The folks he has trained in evidenced-based protocols are then developing new ideas and testing those in evidenced-based treatment.

The goal of Contextual Psychology is to help the client function apart from therapy. The goal of Contextual Psychology is not to have them keep coming back week after week, but to take what they've learned between the sessions, apply it, test it, find out what works most effectively for them in their unique context, and then ultimately to be able to function apart from having to go back to the therapist.

Hayes' research into human emotions began by studying the problem of social anxiety, a very common psychological problem that can be debilitating for many people. Hayes observed that the tools we use to solve problems actually lead us further into traps that create suffering. This is where Relational Frame Theory really fits into the picture. Our ability to create arbitrary meaning is an evolutionary trait in psychological evolution that means humans rule the planet rather than giraffes; however, this otherwise useful tool can actually lead us into further traps that create suffering. One of the basic premises of ACT therapy is that fighting pain actually just makes it worse. That's counterintuitive.

It really doesn't seem to be logical, but the research shows that fighting pain actually makes it worse. Probably one of the simplest explanations is that you have to direct your attention to that which you are fighting. Many of you have of course explored this thought experiment:

Don't think about a yellow jeep. Don't think about its yellow hood, or the yellow doors, or the yellow mirrors. Don't think about how the black tires look against the yellow body of that yellow jeep. Don't think about the chrome bumpers on that yellow jeep.

Now, what are you thinking about? What are you picturing? A yellow jeep, of course.

The idea on ACT therapy is not to become un-depressed or un-anxious or even un-angry, but rather to see our thoughts as just thoughts, our emotions as just emotions. The goal is to help a client to detach from the relational frames that cause so much distress.

Here's the definition of ACT therapy the Association for Contextual Behavioral Sciences offers:

> *ACT is an approach to psychological intervention defined in terms of certain theoretical processes. Not a specific technology.*[17]

In theoretical and process terms, we can define ACT as a psychological intervention, based on modern behavioral psychology including Relational Frame Theory, that applies Mindfulness/acceptance processes and commitment/behavior change processes to the creation of psychological flexibility.

I recently co-wrote a book titled *Counseling People Who Have Killed Other People* with my colleague David Parke. In his work with military veterans and in leadership positions when he was serving in the military, David Parke talked about developing resiliency as one of the keys to mental health. This is part of psychological flexibility, an important aspect of ACT therapy that it makes it particularly useful for a wide variety of different types of clients that we work with.

In *Get Out of Your Mind and Into Your Life*, Steven Hayes points out five psychological truths. They are based on research, yet they are counter-intuitive to the way most of us would think about psychology and emotions.

The first truth is that you cannot deliberately get rid of your psychological pain, but you can take steps to avoid artificially inflating it.

---

17 "The Six Core Processes of Act", Association for Contextual Behavioral Science http://contextualscience.org/the_six_core_processes_of_act (accessed March 8 2014).

Pain and suffering are two distinct states, and by inference they do not need to coexist. You can have pain without suffering. I work with a lot of pain control clients, and at first that idea is difficult for them to grasp. Nonetheless, when they do learn the distinction between pain and suffering, they find it tremendously freeing.

This is true for our clients with emotional pain as well. One of the other truths outlined in *Get Out of Your Mind and Into Your Life* by Steven Hays is that you do not have to become your suffering. One of the most difficult of all the presenting problems in counseling is that our clients become fused with their problems: they have in fact become their anxiety. They have become their anger. They have become their depression. They have become their failure.

Hayes' research at the University of Nevada shows that paradoxically accepting your pain is actually the first step into ridding yourself of suffering. And he points out something else that really flies in the face of everyday thought, and that is psychological pain is normal. It's actually normal to experience psychological pain. He calls it important and says that everybody has psychological pain.

Clients often come to us with the unrealistic belief that other people do not have depression. Other people don't have anxiety. Other people don't have these problems. The reality is everyone has psychological pain. It is in fact normal.

Mindfulness is a key concept in Acceptance and Commitment Therapy because it is a primary way to separate pain from suffering and to take steps to avoid artificially inflating psychological pain. Acceptance is the notion that trying to get out of our pain only amplifies it. Trying to get out of your pain entangles you in it. That's what produces trauma paradoxically. The alternative here is acceptance.

The paradox of quicksand is applicable here. Now, I have never actually been stuck in quicksand, but that they say that if you find yourself in a pool of quicksand, your natural inclination will be to try to pull yourself out of it. However, every time you step up, you create a hole.

The suction underneath pulls you deeper, until you eventually drown in the quicksand.

The only way to actually get out of quicksand is quite counter-intuitive: You must make yourself as still as possible so as to float to the top. Then you can roll yourself out of the quicksand rather than fighting the quicksand. You have to go with the quicksand rather than to go against the quicksand.[18]

In our profession, we have almost no training in modalities that help clients to achieve acceptance. In fact, page 449 of the Big Book of Alcoholics Anonymous is the closest thing we have. It teaches us:

Acceptance is the answer to all my problems today. When I'm disturbed is because I find some person, place, thing or situation, some fact of my life unacceptable to me and I can find no serenity until I accept that person, place, thing or situation as being exactly the way it's supposed to be at this moment. Nothing, absolutely nothing happens in God's world by mistake and until I could accept my alcoholism I could not stay sober. Unless I accept life completely on life's terms I cannot be happy. I need to concentrate not so much on what needs to be changed in the world as on what needs to be changed in me and my attitudes.

The commitment part of Acceptance and Commitment Therapy is of course values and values-based living. To understand values-based living, Hayes poses several questions to the reader, and these are questions we should pose to our clients:

Are you living the life that you want to live right now?

If the answer to that is no for you or for your clients, then the solution is to right now put yourself where you want to be. If you're not happy, put yourself in the state of happiness. It's amazing we actually have control over our own states. I can be happy even if I experience pain. That is something that my clients do not know. It has to be taught to them.

---

18 Hayes.

Is your life focused on what's most meaningful to you? If not, what action can you take today to drive meaning?

Shifting from uselessness to mental management to life engagement is a process of values-based living. It's our mental games, our relational frames, which keep us from experiencing life engagement now.

One of these mental games is the belief that something must be resolved or happen before we can experience or make a change. Those are faulty mental premises.

There are six core processes within Acceptance and Commitment Therapy.[19] This chapter does not present exhaustive coverage of Acceptance and Commitment Therapy, but a brief summary of these core processes will serve as an introduction. These six core processes provide a conceptualization for a complete therapeutic process with a beginning middle, and end.

Core Process One: Acceptance

Clients often present trying to get rid of pain: "I don't want to be anxious anymore. I don't want to be afraid to fly. I don't want to smoke anymore. I don't want to be obese." They often want to be rid of pain, or to forget the past. Too often, they come in trying to change something that is unchangeable.

Acceptance is not about liking something, hoping that it happens to other people, or wishing that it happens again, but simply recognizing its existence. As the quote from Alcoholics Anonymous says, acceptance is the answer to all of my problems today.

Core Process Two: Cognitive Defusion

Clients come to us because their previous strategies have failed. The previous strategies to problem solve have actually made things worse.

---

19 "The Six Core Processes of Act".

By paying attention to the pink elephant in the room or the yellow jeep, they have become fused to their thoughts. They now believe what they think, even though thoughts and facts are often two different things. As a result, they're aware only of the pink elephant in the room or the yellow jeep that they're trying not to think about. Cognitive defusion techniques train your client to no longer be fused with troubling thoughts.

Core Process Three: Being Present

In my sessions, being present is actually one of the first things I teach my clients, because it takes time for them to practice before they really understand what the value of being present.

Core Process Four: Self as Context

This is really about using a strategy to help a person become their own observant self. This is one of the reasons why we focus on the breath. It's to practice observation rather than enmeshment. See, my clients are just angry. They don't observe anger. My clients are just depressed. They feel it. It permeates them. They don't observe depression.

If you have studied Neuro-Linguistic Programming, you may be familiar with first, second, and third perceptual position. The first position is me as me, but in dissociation we can also move to a second perceptual position. This is where I would be an observer seeing me in a situation. In the third perceptual position, I am dissociated even further to where I'm an observer watching an observer watch me. By taking our clients through processes where they step out of themselves and observe themselves as if from above or from a new vantage point, they can really begin to practice seeing themselves in context of a situation rather than as the situation.

Core Process Five: Values

This process is about choosing a direction and establishing willingness. Willingness is really a key word in ACT therapy, and in the next chapter, I'm going to share with you an exercise I have adapted from *Get Out of Your Mind and Into Your Life* called the *Willingness Dial*. This exercise helps your clients to identify willingness and motivating values. Willingness is not resignation. It is also not the same thing as wanting. It is a willingness to experience, to accept in the face negatively-valued emotional states.

Core Process Six: Committed Action through Practice

All of my clients are prepared each session to build on the learnings of the previous session so that when therapy concludes, they can continue to practice these things in a committed action towards a valued path. It sees themselves in context of not just an internal experience but as an outside observer being mindfully present in that world directly—not fused or meshed with their emotion—and accepting that life is sometimes difficult.

These six core processes can become the foundation for a structured program in ACT therapy or classes in ACT therapy; or it can be modified based on an individual in different formats of therapy, hospitalization, partial hospitalization, outpatient therapy, counseling, coaching, or strategic intervention to maximize the potential of our client.

ACT Exercises

Perhaps the best way to understand these processes is to experience them. Now typically, I lead clients with my voice and ask them to close their eyes, but as you are reading, I will lead you through several of these as eyes-open processes. Of course, you can close your eyes after reading each one in order to enhance your imagination, or you can simply experience them with your eyes open. When you work with clients, I recommend having them close their eyes.

To a large extent this is very similar to the process of clinical hypnosis or clinical hypnotherapy. Although I do a fair number of training programs on clinical hypnosis and clinical hypnotherapy, and in my office I practice clinical hypnosis, I'm always amazed at how many mental health professionals are unfamiliar with the modality of hypnosis or hypnotherapy. They even come to my trainings say, "I'm not sure I can learn this."

All therapists already do hypnosis. Albert Ellis wrote and entire chapter on it in his 1970 book, Rational Emotive Behavioral Therapy. It doesn't matter what approach or theoretical orientation you've had to this point. You certainly are using hypnotic modalities of treatement. Because of Hollywood, we think of hypnosis as a magical or mystical state, or we see trance as something that only occurs because of an altered state of consciousness. The reality is there's absolutely nothing, zero that happens in hypnosis or hypnotherapy that doesn't happen in regular life.

If you've seen a stage hypnotist, recognize that it is theater. That's what Hollywood likes to represent rather than what hypnosis really is all about. And every therapist, whether they have ever taken a hypnosis training course or not, does hypnosis.

Think of it this way. I've been into a lot of therapists' offices, and they're almost always set up sort of like a relaxing living room, with a comfortable recliner or a comfortable couch or wing-back chairs with big arms to fully sink into. Whether we've ever taken a hypnosis course or not, when we take somebody outside of the busy world and their enmeshment with the difficulties of life, and then we have them relax in the nice environment that we've provided for them so that they can truly focus on the change that they need to make, that is in fact the essence of the hypnotic process.

Hypnosis is one of those loaded words that people are often little mystified by, sometimes even scared of, but the reality is we all experience hypnosis every single day, and we experience trance every single day. In fact, hypnosis or hypnotherapy isn't about getting somebody into hypnosis or getting them into trance. That's what Hollywood is all about. Hypnotherapy is really about helping our clients to utilize the trance states that naturally exist.

Have you ever watched a movie and cried? Let's take the movie *Marley and Me* for example. You cried at the end when the dog died—but they didn't really kill a dog in order to make the movie for your entertainment. No dog actually died during the making of the movie *Marley and Me.*

There are actually fourteen dogs used in that movie and quite a few children because the kids keep getting older, too. So, nobody was actually harmed in the movie *Marley and Me,* and no dog was actually killed. Yet you really did cry when you saw that movie. There was no dry eye in the theater when I saw that movie.

That's really an example of hypnosis or experiential theater as Don Gibbons calls it where we have suspended that critical faculty of the mind and fully engaged in the experience in this case of a Hollywood movie to produce real emotion. As therapists, we all do this in our offices. We engage clients in an imaginary process that produces real change. So when I call these processes hypnotic, we should recognize that this is something all of us do—even Albert Ellis did it.

In fact, a lot of people forget that Albert Ellis, in his seminal book *Rational Emotive Therapy,* which was the basic textbook for Cognitive Behavioral Therapy in the 1970s, devoted a whole chapter specifically to the value of utilizing these trance states. And of course, Milton Erickson was a family therapist as well as a psychiatrist who used language patterns to effect change in his clients, and those approaches are certainly hypnotic methods as well.

So, these three exercises that I'm going to guide you through are experiential processes. I use these with my clients sometimes in a formal process of hypnosis, but generally without such a formal process. I'll be talking to my client, interviewing my client, discussing with them their valued path, discussing with them their thoughts, their feelings or experiences, and then I'll guide them through a process:

Finding Your Valued Path

Imagine that you've just received a present. Imagine that present is a surprise. You don't even know who sent it, but you are now holding it. It won't be a simple gift, but rather a magical gift, the one thing that you want more than anything else in life. Maybe it's big or perhaps small or maybe it's too big to actually hold. It's your present, so you can see it any way you want to. You can even hear the present.

When a child is trying to figure out what's inside of a gift, they often shake it. You can do that in your own mind, hearing the gift inside of the box. Feel the shape of the box. Is it square, rectangle or even an odd shape? Of course, this is your gift, I don't want to hold you back, so at this moment you can begin to open the gift using that creative imagination, finding inside the wrapping exactly what you wanted more than anything else.

Perhaps it's the deed to a vast and beautiful land or a diamond worth millions of dollars or even a winning lottery ticket. It's your gift. You can create it as you want it, so go large taking this opportunity to discover what you would want more than anything if you are to receive a magical gift. Allow yourself to dream big, to open the present and enjoy that magnificent gift inside. Now, the papers that you opened, wrapping papers are lying to the side. You're holding that unwrapped gift in your hand or place the gift out in front of you and just breathe.

See your gift and imagine all the changes in life as a result of getting that gift. And note how life will change because you got that gift. How will you feel? What experience will it bring? Perhaps a sense of freedom, maybe a respect from other people. Will you be able to help others now that you have received this gift? Will the gift help you to be more secure or even to be more important? Take a moment and explore this idea in your mind, the idea that you have this gift and some of your deepest needs can be met.

What are those deepest needs? Become aware of them. Note the words that come into your mind that describe these needs and notice the feeling that you associate with having imagined receiving this gift. Take in a breath. Feel fantastic. Think of two or three words that describe the gift, and say them aloud.

It's interesting how what we truly value or need is often represented by an object or something tangible, in this case the gift. What was actually inside of that box? But with little reflection you soon discovered what is most important is not the gift itself, but those words that you just said what the gift represents. You described what you valued most as blank and blank. This exercise has helped you to identify your core values. This is your valued direction.

Of course, you can use this with your clients, and you will adapt and tailor it to each client: the speed, the rate, what it is that you focus on, and how you follow up with it. But this is really a great exercise for helping clients use the resource states that are within them to identify those core values that can help them to identify a valued direction. When you have clients who are stuck in therapy, this is a great exercise. I've done this with couples where Bob and Bertha are sitting next to each other, both with their eyes open, both coming up with different gifts, both coming up with different words, not even processing with them the exercise, but simply taking notes to determine what our valued path in couples counseling is going to be.

Willingness Dial

Remember when you were a kid, and you might have had one of those toys called The Chinese Finger Trap? It was toy many people played with; it was made out of straw braided together into a tube. And when you put your index fingers of each hand into the ends of the tube, it constricted in such a way that when you try to pull them out, it created a virtual finger cuff. It became impossible to remove them by pulling them out.

The harder you try to pull out, the tighter the braids on that straw became. Of course, in order to finally get out of the finger trap— the way to finally free the fingers—was to stop pulling to break free, to stop resisting the trap. That's counterintuitive to say the least, which is why the Chinese finger trap toy was not used as a child's toy, but an ancient torture devices. The only way to find freedom from Chinese finger trap was to let go of the struggle.

Years ago, there was a song by a Presbyterian minister named John Fisher and the lyrics of the song said, "Losing is winning when it turns you around. It all looks clear with your feet on the ground. You walk out a winner."

One of the great paradoxes of the mind is that the more we struggle to solve a problem, the more constricting it becomes. And the true freedom often comes from submission. And this of course brings up the subject of willingness and you must ask yourself, how willing am I to stop resisting the struggles? How willing am I to stop resisting depression, anger, anxiety?

Willingness to change and commit to a valued direction is really where you are right now. Of course, by being willing to stop to struggle, it doesn't mean that you will lose. To the contrary, it means you'll give up the suffering—and that's where you will find freedom. So, I'm going to ask you. How willing are you to give up control of your internal thoughts and feelings? How willing are you to just let fear be fear, to let sadness be sadness, without struggling to stop it, control it, or direct it?

Imagine in front of you a dial with 1 on the low end and 100 on the high end. To what degree, essentially to what number are you willing to give up control? Think about that number now. To what number on that dial--1 at the low end, 100 at the high end—are you willing to give up trying to control your husband, your wife, your anxiety, your fear, your drinking, whatever it is? If you chose a low number, you are holding the idea that a low number means that you'll experience much less pain.

But of course, that's not what experience has taught you. If that were true, the resistance and thought suppression was effective, you wouldn't be here. So, consider your score and your willingness to increase the number, from a 5 to a 10, from a 10 to a 20, from a 20 to a 40, from a 40 to a 50, even for a 50 to an 80 to 90 or even maybe 100 being willing to give up trying to control that drinking spouse, fear, depression, and find that when you increase that number, you increase the acceptance of pain.

Paradoxically, your suffering can be released forever. And so now I'm going to ask you, at what number from a scale of 1 to 10 are you willing to stop resisting and to simply accept.

It is at this point if I'm with a real client that I'm probably going to be bridging into some real issues for the therapeutic process. This is probably going to be one of those breakthrough points in this particular session where I use a process like the willingness dial.

This is really incremental therapy. Instead of moving from here to here, we're simply moving from here to a willingness to be somewhere else. In the movie *What About Bob*, Richard Dreyfuss plays a psychiatrist. Bill Murray plays his client, and the fictional book that the psychiatrist

Richard Dreyfuss wrote was called *Baby Steps*. It was a great movie. It actually taught something that was actually true: we don't need to make all of the changes today. Today only requires willingness. When we achieve willingness, we'll move to where we need to be.

Looking in the Box

This process of understanding and experiencing acceptance uses the imagination. And so imagine that you're in a large conference room. All the chairs surround the large table, and you're sitting in the executive chair at the end. All of the other chairs in the conference room are unoccupied. Imagine you're the only one in the room.

A large box sits in the middle of the table. The box on the table of course holds all of those things that you've not wanted to look at or see, whether because of pain or fear or for any other reason. But you are safe, not only in the conference center that you've created, but in this actual office here.

You've come here to move forward, and so this is actually your opportunity in the safety of this office to begin unpacking these things from the box and placing them on the table. So step up to the box and open the box and one by one unpack those items from the box, looking at them without judgment or attachment simply acknowledging what those things are and placing them on the table. See yourself doing this or hear the sound of those items landing on the desk as you take each item out and simply place the items from this box on the table.

There may be only one or two things or maybe five or six things. Maybe there are many things. But when the box is unpacked, look at all of those items. See them, but rather than sitting again, simply stand at the end of the table, taking as much time as you need in silence to simply see those things out on the table.

[If it's too painful at this point, you can always have a client move into a second perceptual position, being an observer, possibly from the other side of the glass seeing themselves looking at those things on the table.]

Of course, you're safe. You're the only person in the room observing the table. When you've looked long enough at all of the items on the table, imagine you approach the doorway leaving those items exposed on the table not packing them up again. Turn out the light. Walk through the door, closing it behind you. Those things, feelings and experiences are neither hidden nor neglected, but as you walk out the door and into the light of this new chapter of life recognize that they were there and that by living fully in the present you've developed a sense of acceptance, simply seeing life as it is and practicing the new art of living fully in the moment.

Deliteralization

There are many different ways in ACT therapy that we can defuse from our unhealthy thoughts. Deliteralization is particularly useful as well as silly and fun. What it does is separate a word from its meaning, which can be a real powerful change.

Suppose for example you have a client who's enmeshed with hurt, the pain of the hurt is so deep they are a hurt person, and that is almost at this point their identity. They literally own the word hurt. It happens in couples counseling all the time. It happens when I'm counseling children of alcoholics, whether adults, adolescents, and even young children.

Their need is to defuse from that word, to put some space between them and that word because that word has become so powerful to them.

What you're going to do is you ask the client to say the word that they are fused with out loud, mindfully observing the word, saying it as quickly as they can, but still able to enunciate it and pronounce the word. This process is a silly process, but it's an amazing process. Give your client 45 seconds to do this, never less than 45 seconds, never more than 60 seconds. When they're done with the exercise, ask them to really check if by now saying that word feels different than it did a few moments ago in any way, shape, or form. Your client will almost always say yes, as they have put some space between themselves and the word.

Don't just read about this process. Try it yourself and experience the change.

By the way, try this exercise in deliteralization in couples counseling with the name of the mistress or the mister that's been part of the enmeshment of the wounded or hurt spouse. That is a powerful, a powerful strategy in couples counseling.

Changing Self-Talk by Using Submodalities

Albert Ellis taught us that damaging self-talk can drag us down emotionally as we repeat cognitive errors to ourselves. Of course, positive affirmations that counter those cognitive errors can be helpful to many clients, but when our clients are too caught up in their own negative beliefs, positive affirmations can be too great a stretch and can actually reinforce our clients' depression, our clients' anxiety.[20]

So the next two methods are ideal for clients who aren't ready for positive counters to negative self-talk, instead providing defusion from those cognitive errors. In other words, we remove the power of a thought or emotion.

Slow It Down

You've said you feel like you're not a worthwhile person. Usually when you say that to yourself, you just hear that voice in your mind maybe even screaming it or shouting it at you. Now you're going to practice changing that voice and all of those self defeating errors and putting some space between you and the thought.

20 J. V. Wood, W. Q. Perunovic, and J. W. Lee, "Positive Self-Statements: Power for Some, Peril for Others," *Psychol Sci* 20, no. 7 (2009).

Either out loud or in your own mind, say, "I am not a worthwhile person," but this time say it softly and slowly, not as you usually hear it in your mind.

This is a great counter to that negative self-talk that Albert Ellis talked about. Did you notice you responded to that familiar self-talk a little bit differently by making your voice softer and by slowing it down? Try it again this time even slower. Very slowly say, "I am not a worthwhile person." In fact, say it again even slower, almost so slow that it doesn't make sense anymore. "I am not a worthwhile person." Now, ask yourself, how is my experience of this thought different now than it has been in the past?

Of course, it is—because you've created some space between you and the thought. You've changed the submodalities of it, and you're no longer going to automatically respond to that negative self-talk with impulsive choices, but this time you will respond when noting these thoughts by slowing them down and changing the experience of the thoughts.

Negative News Channel

Another idea you can do is to imagine the negative thoughts in your head being broadcast from a Negative News Channel.

You can imagine that this Negative News Channel is broadcasting from your mind. It's even okay to have a little bit of fun with this, to even smile as you hear your negative self-talk through the NNC, the Negative News Channel. Go ahead. Start thinking of what the NNC might broadcast tonight: "This is the NNC, Negative News Channel broadcasting 24/7, 365 days a year. Today, I'm fat, I'm ugly, and I'm a bad person." Go ahead. Identify some of your own negative thoughts and hear them through the voice of the announcer of the NNC:

"I'm stuck in this situation. I can't do anything right. I'm going to screw up the future." Go ahead. Hear it again even in a radio announcer's voice. Listen to the Bad News Radio. It's absurd, isn't it? See, in the past, you automatically bought into the thoughts that you created, but creating the absurd NNC, you heard them played through another person's voice and can now recognize those thoughts as just thoughts, about as meaningless to you as the real nightly news.

Of course, all of these exercises have to be adapted to your personality, to your particular clients, your particular situation. My hope is that you can find ways to creatively adapt to the scripts that have been provided. For example, a practitioner I knew had a Catholic client who found that she couldn't allow her negative self-talk to sound silly in her head, but she could find relief and distance by imagining a cathedral choir singing her negative self-talk; in that way, she could release it to God.

CHAPTER SIX

Positive Psychology

Over the past decade or so, Positive Psychology has been researched from an academic perspective allowing it to move into the category of empirically-based treatment approaches.[21]

Martin Seligman, past president of the American Psychological Association and one of the chief proponents of Positive Psychology, defines the approach in this way:

We believe that a psychology of positive human functioning will arise which achieves a scientific understanding and effective interventions to build thriving individuals, families, and communities.[22]

Positive Psychology is really the antithesis of our traditional approaches, which have a pathological orientation, looking at what is wrong with our clients. In contrast, Positive Psychology looks at what is right rather than what is wrong. It is primarily concerned with researching psycho-

21 M. E. Seligman et al., "Positive Psychology Progress: Empirical Validation of Interventions," *Am Psychol* 60, no. 5 (2005).
22 M. E. Seligman and M. Csikszentmihalyi, "Positive Psychology. An Introduction," *Am Psychol* 55, no. 1 (2000).

logical theory and intervention techniques to understand the positive, adaptive, creative, and emotionally-fulfilling aspects of human behavior.

In other words, traditional approaches have tended to say, "Oh my gosh what's wrong with you? Look at all of these horrible things that have happened." Positive Psychology says, "Wow, what have you done to survive? It's remarkable that you've made it in light of all of these difficulties to the point that you have." It really is a complete change of pace from the traditional problem-oriented model of psychology.

I remember back in the late 1980s learning what a psychosocial evaluation was and writing my first psychosocial evaluation. Back then, I'd write twenty or twenty-five pages. That was before computers, so I had to write most of it by hand. (I was, and still am, a terrible typist). So after spending two hours interviewing a new client for inpatient treatment, I would write page after page about all of the problems they had: their legal problems, political problems, health problems, family problems, sexual problems, emotional problems, problems, problems, and more problems.

The last question on the psychosocial evaluation was always, "What are the client's strengths?" By the time we got to the last question on the psychosocial evaluation, I would be so tired of writing and so tired of interviewing that I'd write something just to fill in the blank: "They have cool shoes" or "They have nice hair" and then I'd move on.

Then one day it occurred to me that this was actually the most important question on the psychosocial evaluation. So I moved that question from the bottom of the evaluation to the top. It became the first question.

Over the years, it's really become the only question that I concern myself with. The intake form in my office is very short. It's about two pages long in really big type, so the intake form really doesn't have very many questions. The reason why is because I'm really uninterested in what's wrong with my clients. I want to know what's right with them, because I use an approach based on the methods of Positive Psychology articulated by Seligman.

Positive Psychology is not, by the way, Norman Vincent Peale's Power of Positive Thinking. There's nothing wrong with that, but Positive Psychology is something different. It is concerned with psychological theory, research, and intervention—finding those techniques which are adaptive, positive, creative, and emotionally fulfilling, rather than simply employing positive platitudes.

Likewise, Positive Psychology is not about the Law of Attraction or positive confession. It's not about thinking only positive thoughts and thereby attracting positive things. That's not what Positive Psychology is all about. It is not a fad.

Positive Psychology is not about ignoring the existence of problems; rather it's about recognizing that some of our clients' problems are so catastrophic there's probably zero that can be done in the context of therapy. For most of my clients, those problems are a result of their histories—and history can never be changed.

Positive Psychology is not a simplistic approach. There are volumes of resources that have written. One of the best is the *Handbook of Positive Psychology,* which includes hundreds and hundreds of pages of techniques, ideas, and strategies, all based on solid research.

It is a non-pathological model or approach not unlike Person In Environment (PIE). It identifies the strengths or resources. It recognizes that it's a lot easier to strengthen the resources that exist within a client than it is to create new strategies that to this point have not yet existed in almost every one of my clients. In fact, I can honestly say in all of the clients I've ever worked with, although many of them have had tremendous problems, I've never met anybody yet who had no strengths and no resources to draw from.

They may have been limited. They might not have known how to use them, but every client I've ever worked with has had strengths and resources available to them. Positive Psychology is a set of strategies that can be applied both to individuals and to families as well as orga-nizations.

The process known as Appreciative Inquiry, which is an outgrowth of Positive Psychology, has been used with community service organi-zations as well as huge multinational corporations. It is an application of Positive Psychology to community development, very useful for social workers, ministers, corporate consultants, and organizational psychologists.

A small social service agency in Phoenix, Arizona, called Neighbor-hood Ministries, has used Appreciative Inquiry as a process for devel-oping community over the last 30 years with tremendous success and efficacy in their work.

After September 11th, British Airways used Appreciative Inquiry as a process for changing their organizational culture. They experienced financial success in a time when many airlines, particularly international airlines, were having tremendous difficulty.[23]

People are familiar with the corporate story of Jack Welch and General Electric will know that Appreciative Inquiry was used as a con-sulting process with all of the employees of General Electric during Jack Welch's tenure.[24]

So, Positive Psychology has applications far beyond individual therapy.

Positive Psychology is a non-threatening approach. It doesn't seek to embarrass or humiliate by exposing what is vulnerable and what is wrong, but instead by valuing what is right. It is motivational, so it is a great thing for those therapists who are following a coaching model in that ultimately it seeks the highest level of performance.

---

23 Jane Magruder Watkins and Bernard J. Mohr, *Appreciative Inquiry : Change at the Speed of Imagination*, Practicing Organization Development Series (San Francisco, Calif.: Jossey-Bass/Pfeiffer, 2001).

24 Ibid.

The word often associated with Positive Psychology is *flow*—and flow is being in that zone of happiness. A lot of happiness research has been done in the last several years by those interested in Positive Psychology. Flow is also described as that state of absorption in which one's abilities are well matched to the demands at hand.

Positive Psychology is really concerned with three main issues: Positive emotions, positive individual traits, and of course, positive institutions. There is an acronym for the goals of Positive Psychology as articulated by Martin Seligman: PERMA.

- P stands for Positive Emotions—feeling good, feeling happy, feeling a sense of flow.

- E stands for Engagement. Our goal in therapy is to help our clients to become completely absorbed in the activities that promote flow or health or wellness.

- R stands for Relationships. While the Internet through social media can give us connections to perhaps more people in the current era than we ever could have connected to in the past, many of our clients still lack authentic connection to others. The techniques of Positive Psychology are designed to support that goal.

- M stands for Meaning. Just as Steven Hayes emphasizes a valued path, Seligman discusses purposeful existence.

- A stands for Achievement. This is really about a sense of accomplishment and success. We all need to feel that we are valued, that we have accomplished something, that we are successful. Think about the important life events that you've had. Those life events are probably things that have been related to your own personal success or an accomplishment: building a new home, graduating from college, doing whatever it is that you did that brought you a sense of achievement. Have you ever won an award for anything? We value that sense of accomplishment and success. It's not enough to simply make change without deriving a sense of accomplishment and success from that change.

The hallmarks of Positive Psychology include self-efficacy. This is really the belief in one's own ability to accomplish a task. There are a number of strategies that we can use with clients to help them create a sense of self efficacy, a sense that not only have they accomplished something, but it was through their own efforts that they've reached their goals.

One of the other hallmarks of Positive Psychology is personal effectiveness. This is primarily concerned with planning and implementing methods of accomplishing those things that are important to them.

Flourishing is another hallmark of Positive Psychology. Although the concept is tied to Seligman, Fredrickson gives us a great definition:[25]

> *It is optimal human functioning. It comprises four parts:*
> *Goodness, creativity, growth, and resilience.*

Now, think about the counseling that you've been doing over the last years in your office, when clients are with you--Is it focused on goodness? Is it focused on creativity? Is it focused on growth? Is it focused on resilience?

Or has much of that counseling been negative? We're trying to solve, but the main focus has been on what's wrong. Positive Psychology helps us to flourish in therapy by focusing on optimal human functioning rather than what has been wrong.

Yet another key element of Positive Psychology, as in any other Contextual Psychology, is Mindfulness—intentional focused awareness on one's immediate experience.

Elevation, another important element of Positive Psychology, is

---

25   B. L. Fredrickson and M. F. Losada, "Positive Affect and the Complex Dynamics of Human Flourishing," *Am Psychol* 60, no. 7 (2005).

defined as that pleasant moral emotion involving the desire to act morally and to do good. For those readers familiar with Kohlberg's Stages of Moral Development, Elevation is that highest level of moral development. So that we can achieve our highest level of potential, Elevation is something that we strive to accomplish in Positive Psychology. That's really different than most approaches in therapy, which in the current managed-care era of limited time and limited resources, simply try to help people move from a pre-crisis level of functioning back to an adequate level of functioning.

By employing approaches of Positive Psychology, we can move people to Elevation. This is important not only to our clients, but to our own survival as professionals. When you are home at the end of the day having elevated people rather than simply supported them, it makes a huge difference in your feeling of your own self efficacy, personal effectiveness, flow, flourishing, and mindful awareness.

To accomplish these things, Positive Psychology really asks the question differently than traditional therapy. In traditional therapy, we usually ask the question how can the client become less depressed or how can the client become less anxious or how can the client become less angry. In Positive Psychology, we say although the client is angry, he's experienced happiness, so let's do the opposite of anger. Although the client is angry, he's experienced serenity before. The question is not how can we stop anger, but how can we increase serenity.

So, the therapeutic approaches are about increasing serenity, not about decreasing anger. Instead of asking our client what can you do so that you're not depressed, we will ask our client instead what can you do to feel happy. It really switches out the questions and has them operate from the other side of the pool table.

Some of the specific methods of Positive Psychology include of course teaching our clients positive thoughts, how to reframe thoughts, how to find that which is positive.

One of my favorite approaches helps my depressed clients who say, "I can't be happy."

I will say that my client, "You know, a lot of people in their house have a photo album, or maybe a baby book, or a family photo album or something like that—could you bring that in here next session?" And then when they come in to that next session, I go through the pictures with them. There might not be many pictures where they look happy, but I have always found at least one picture where a client looks happy, so I'll say, "It's amazing—in this picture, you look happy. What was going on then?"

And they'll say, "You know what, that was the last time I was happy. It was 28 years ago, but it was on that day because this is what happened. This is what happened. This is what happened."

I point out, "Now, we know that even though you're not happy, you have the capacity to be happy. We know that because you'd been happy in the past. So now the question is not why is it that I can't ever be happy, but how was it that I can increase happiness that I've felt before."

Affirmations

Back in the 1980s, Stewart Smalley, a character on Saturday Night Live, used to sit in front of the mirror and say, "I'm good enough, I'm smart enough, and doggone it people like me." That was his affirmation to himself. It was a Saturday Night Live skit that really spoofed self-help programs and the whole affirmation movement. I actually love affirmations. I think that they can be a tool in Positive Psychology. I think that Stewart Smalley may have been funny, but Stewart Smalley actually got it correct.

I teach my clients the art of using affirmations. Affirmations, when used well, are extremely effective. The reason why is simple: Anybody who's been practicing Cognitive Behavioral Therapy knows that we need to counter cognitive errors. The most effective way to counter cognitive errors is with the truth, and those truths can become affirmations.

Outside of any Twelve-Step meeting, you'll see cars with bumper stickers reading, "Just for Today," or "One Day at a Time." Those are affirmations. Those affirmations have become one of the most important elements of the success of Twelve-Step programs. Why? Because they are effective counters to cognitive errors.

At the current time, our cultural awareness of the value of affirmations has moved beyond Stewart Smalley to the Law of Attraction in *The Secret*. You may recognize some of these phrases: *Like attracts like* or *birds of a feather flock together* or this one, *it is done unto you as you believe*. These are examples of affirmations, and there seems to be in pop culture right now renewed interest in affirmations, probably in large part because of *The Secret*.

Now, at the beginning of this chapter I wrote that Positive Psychology is not the Law of Attraction or *The Secret*. But because our pop psychology culture is aware of that movement, when we ask our clients to learn skills related to affirmations, they are going to relate it to *The Secret*. It's okay if our client comes to us with a set of experiences or knowledge or beliefs to actually utilize that resource. Of course, there are a lot of clients who've never heard of *The Secret*.

Similarly, a lot of my clients are familiar with the power of the Word from their study of the biblical creation account. It begins of course in Genesis with God speaking, and this creation story is reflected both in Christian tradition as well as the stories of other cultures. In Buddhism, one aspect of the Eightfold Path called the Sama Vaca is often translated to English as Right Speech. The more accurate rendering is Wise Speech or even Skillful Speech, which hopefully suggests speech that is acquired through practice.

The power of affirmations

Not only does our culture, whether in Saturday Night Live, The Secret, or religious scriptures, talk about the power of the spoken word, philosophers and metaphysicians in the past hundred years have written profoundly on this subject. Charles Haanel, who is an industrialist and the founder of the St. Louis Post Dispatch Newspaper, wrote these words:

> *Thought is energy. Active thought is active energy. Concentrated thought is a concentrated energy. Thought concentrated on a definite purpose becomes power.*[26]

A century before *The Secret* movie was ever produced, Charles Haanel really understood the essence of this aspect of the spoken word.

William Walker Atkinson, a metaphysician and philosopher from the early 1900s, wrote the best way to overcome undesirable or negative thoughts and feelings is to cultivate the positive ones.[27]

Much more recently, Sports Psychology Journal in 1991 published an article that was fascinating. They found that healing from injury most significantly correlated to goal-setting, positive self-talk, and the use of healing imagery.[28] In the book I wrote a few years back titled *Medical Meditation*, I discussed the techniques of goal-setting, positive self-talk, and healing imagery as aspects of promoting wellness whether it was a sports injury or any other kind of injury.

---

26 Charles F. Haanel, *The Master Key System in Twenty-Four Parts with Questionnaire and Glossary* (Saint Louis, Mo.: 1919).
27 William Walker Atkinson and Harry Houdini Collection (Library of Congress), *Thought Vibration, or, the Law of Attraction in the Thought World* (Chicago, U.S.A.: Library Shelf, 1910).
28 L. Ievleva and T. Orlick, "Mental Links to Enhanced Healing: An Exploratory Study," *Sports Psychologist* 5, no. 1 (1991).

Where do we come up with affirmations?

When we listen to our clients, they will tell us what affirmations they need to make. On my intake form, I ask clients to list three strengths that they have, and I'll use those strengths as a basis for eliciting affirmations.

You're going to elicit affirmations from your clients by reversing their deficits or their presenting problems.

If you are familiar with the Miracle Question from Solution-Focused Brief Therapy, which will be covered later in this book, the answers to the Miracle Question provide fantastic ideas for affirmations and positive visualizations.

When I'm talking to my clients, I help them elicit the affirmations rather than giving them their positive affirmations. I encourage my clients to write down their affirmations so as to use the five Ps:

- present tense
- positive
- personal
- proactive
- passionate

How Do We Use Affirmations?

Of course, affirmations can be repeated daily, like a rosary, but there are many other ways to use affirmations.

None of my clients ever leave my office without this set of instructions: I give them a dry-erase marker, saying take this dry-erase marker home and write this positive affirmation, this truth on the bathroom mirror. That way, the first thing they see in the morning, the last thing they see at night, and something they see repeatedly throughout the day is the counter to their cognitive error, staring back at them in their own handwriting.

This can become a powerful technique, a simple method for helping our clients make change. If you don't have a drawer full of dry-erase markers, go you know buy them in bulk from an office supply store. I guarantee you even if they don't believe it when they wrote it, they will test the affirmation to find if it's true. It will almost always results in dramatic change.

I also ask my clients to put them on a screen saver on their computer. You can create your own screen saver in about two minutes nowadays. Affirmations can be shared as a Facebook status. It's amazing how the social reinforcement, the likes that a client receives in response to posting a positive affirmation, can make that affirmation come alive. Interestingly enough, there's actually a ton of research showing that that's true.[29] The *likes* are a very important part of psychological phenomena.

Affirmations can be recorded by my clients as an MP3, something they can use as part of a meditation or really any other aspect of repetition.

On the topic of affirmations, I should add that a few studies suggest that positive affirmations can exacerbate depression or anxiety.[30] It's important to recognize the need to take baby steps—an affirmation that conflicts too greatly with your client's beliefs can simply lead to self-talk that reinforces the negative belief. However, this does not mean affirmations should be abandoned as a tool, but rather that we should use them wisely.

---

29 C. L. Toma and J. T. Hancock, "Self-Affirmation Underlies Facebook Use," *Pers Soc Psychol Bull* 39, no. 3 (2013).
30 Wood et al., "Positive Self-Statements: Power for Some, Peril for Others."

APE Method

Positive Psychology provides an approach to questioning with clients where we're going to search for alternatives. The acronym for this method is APE:

- ◆ Accuracy: A more accurate way of seeing this is _____

- ◆ Perspective: The most likely thing to happen is _____

- ◆ Evidence: I know this will happen because _____

This technique helps us to challenge and confront those cognitive errors, taking the negatives and asking them to be reframed as positives.

Scaling and Changing Emotions

Scaling the emotions into a Likert scale can be an effective way to help your clients develop positive emotions. Almost all my clients come in and say they're either angry or depressed or anxious. They've never learned that they're actually anxious at a level 8 or depressed at a level 6 or angry at a level 7.

The Likert scale is a scale that measures equal intervals on a scale of 1 to 10. We're all familiar with the Likert scale even if we didn't know its name. If you've ever had a client who makes a mountain out of a molehill or makes a molehill out of something that ought to be mountain, then scaling emotions into a Likert scale can be a real useful tool.

Another idea for helping our clients to develop positive emotions is to help them become aware of co-existing emotions. When they tell us how they feel, ask what else they feel. Ask, "Is there another word that describes how you feel?" Ask your clients to identify coexisting emotional awarenesses. Many of our clients are actually unaware that depression and happiness can coexist or that anxiety and acceptance can coexist.

When we ask them to practice identifying coexisting emotional awarenesses, our clients easily see that while they may in fact be depressed, they were missing that part of them that is happy. One of the primary methods for helping my client become skilled at doing that is to help them learn adjectives that describe human emotions. I find that many of my clients have a limited repertoire of words to describe how it is that they feel.

The fourth idea for scaling emotions or developing positive emotion is really a technique from Neuro Linguistic Programming called a State Control Pattern. What I'm talking about here is not Arkansas, Missouri, or New York—what I'm talking about in regards to NLP states are the experience of an emotion. It's more than just a descriptor of an emotion, but a praxeological vantage point of what it is that they're feeling at that time.

See there's something important my clients don't know: They don't know that they actually have the same pants to get glad in, that they have to get mad in. So I literally practice with my clients helping them to develop a state, a resource state, in my office.

For example, if the resource state is serenity, I'll take five minutes with them and help them become aware of serenity, amplify that sense of serenity, increase the feeling of serenity or whatever positive emotion it is. I then teach my client that if they were able to create that here in my office, they can actually take it to go and they can recreate that anywhere they like to. You can watch a YouTube lecture I created on this subject and these techniques here: http://youtu.be/EQTwPaTyiDA

Seligman gives us an exercise called Three Good Things.[31] It's really an exercise in journalism. He asks his clients each night before they go to sleep to cultivate the practice of identifying those things which are positive. He's say, "Think of three things, three good things that happened today. Write them down in your journal and then a paragraph

---

31 Seligman et al., "Positive Psychology Progress: Empirical Validation of Interventions."

or two reflecting on why did those good things happen." What we have here when a client keeps a journal like this is that at the end of thirty or forty-five days, our client actually has in their own hand their own recipe book, strategies to help them create good things. These become the resources that have come from within the client rather than things that I've given to the client.

This practice creates a habit of doing good things. One of our goals in Positive Psychology is to teach our clients that they can accomplish something and because they can accomplish this, they can accomplish that.

Baumeister actually tells us if you can do anything that requires self-regulation then that makes it easier for you to have self-regulation in everything else.[32] So, for example, suppose you have a client who comes in and says, "I need help. I am a pathological gambler. I have no ability to self-regulate. I go down to the casino every time I get paid, and I blow all my money." Or suppose you have somebody who comes in and says, "I'm a sex addict, and my marriage is over. My life is out of control. I've been impulsive sexually, and I don't seem to be able to stop no matter what I do." The basic psychological principle that Baumeister argues is that when we can develop self-regulation in one area, then like a muscle our self-regulation becomes stronger; it becomes something that's strong enough to apply in any situation.

And so it turns out that our grandmother was actually correct: Sit with correct posture. Hold your head up. Put your shoulders back. Hold your tummy in. Good posture. It's important. But of course, good posture is one of those things that takes self-regulation.

So, when I have a client who comes in with a gambling compulsivity, sexual compulsivity, or really any other of the impulse disorders ranging from trichotillomania to kleptomania, I might actually work with them during the session to cultivate a positive habit, to learn how to self-

---

32 Roy F. Baumeister and John Tierney, *Willpower : Rediscovering the Greatest Human Strength* (New York: Penguin Press, 2011).

regulate something as mundane as posture. If the client can master that, they've really worked out that self-regulation muscle, and they'll be able to apply the strength of that muscle either consciously or subconsciously to any other area of life where they may have found themselves experiencing difficulty.

Further Reading:

Flow: The Psychology of Optimal Experience by Mihaly Csikszentmihaly

Flourish: A Visionary New Understanding of Happiness and Well-Being by Martin Seligman

CHAPTER SEVEN

Solution-Focused Brief Therapy

Solution-Focused Brief Therapy (SFBT) is actually one of my favorite modalities. It is something that I probably incorporate at some level into almost every session I do with my clients—whether it's through the assumptions of Solution-Focused Brief Therapy or aspects of the role of the therapist or the questioning approach that we can use particularly during the assessment process. Of course, like the other contextual approaches presented in this book, Solution-Focused Brief Therapy is an evidence-based protocol with many empirical studies that demonstrate its efficacy.[33]

Solution-Focused Brief Therapy, which is fairly similar to the philosophy of Positive Psychology, has much to recommend it:

- It is a strength-based approach.

- It focuses on solution-building rather than problem-solving.

33 C. Bond et al., "Practitioner Review: The Effectiveness of Solution Focused Brief Therapy with Children and Families: A Systematic and Critical Evaluation of the Literature from 1990-2010," *J Child Psychol Psychiatry* 54, no. 7 (2013).

- It is a competency model that minimizes emphasis on problems of the past and instead highlights the client's strengths and prior successes.

- It operates under the assumption that our clients have had success in something previous to being in our office, and by applying the skill sets and the experiences that they have already had to the present scenario, they're going to be far more likely to succeed.

- It is founded on the grounds that there are exceptions to every problem.

- It teaches that by exploring the exceptions and having a clear picture of the desired future, solutions can be generated by the client and therapist working together as a team.

Here's an interesting story I ran across just the other day about the George Bush Intercontinental Airport in Houston. It had a lot of complaints about the wait for baggage. At first, they added staff. They took other measures to speed things up, and they even succeeded in reducing wait time by eight minutes.

Even though they took these expensive measures, the complaints persisted. So what they did was they actually moved the baggage claim further away from the terminal, resulting in a longer walk to the baggage. It took people longer to get to the baggage claim, but it actually reduced the complaints to almost zero. People spent more time walking, which meant their bags were waiting for them when they arrived.

That's a phenomenal story for Solution-Focused Brief Therapy because what we're looking for here are the solutions that really are outside of the box. Ultimately, we are looking for what works.

There are a number of basic assumptions or a world view of Solution-Focused Brief Therapy, beginning with the idea that therapy is briefer when focused on strengths and solutions.

In the era of managed care, limited time, and limited resources, we're really looking for the shortest path to success. We're trying to limit cost to decrease hospital stays to decrease the weeks in outpatient therapy. This makes Solution-Focused Brief Therapy especially pertinent today.

In the late 1980s, a text book by Gerard Egan called *The Skilled Helper* was used in many graduate school programs. Now Gerard Egan was not known as a solution-focused brief therapist, but his ideas align with this approach. In the book, he proposed a three-scene counseling process, involving the present scene, the preferred scene and the programmatic scene.[34] Each of those three scenes had three stages, so it was really a nine-step counseling process. In other words, what are you going to do to get from where you are now to where you'd like to be?

The first of his steps in the counseling process was, "So, tell me your story." What that means is that we don't ignore the past, but we put it in the perspective: It is one-ninth of the counseling process. It is simply the starting point for our clients, how they got to our office. The bulk of the process, eight-ninths of the process, should be focused on the future.

In Solution-Focused Brief Therapy, we believe the clients have a wealth of resources and solution behaviors that are already present. To a large extent, this is a carryover from the Rogerian idea that our clients have within them all the resources they need to resolve problems.

There is of course the exception to every problem, and in Solution-Focused Brief Therapy we are looking for these exceptions. There are times when a problem wasn't there or a problem was less troublesome for a client; in evaluating those experiences, we can find solutions for our clients.

One of the other assumptions is that therapists help clients find solutions to problems. That's our job. Our job is to really dig in the earth with the client to help them find solutions to problems. We help them

---

34 Gerard Egan, *The Skilled Helper : A Problem-Management and Opportunity-Development Approach to Helping*, 10th ed. (Belmont, Calif.: Brooks/Cole, Cengage Learning, 2014).

set small goals, knowing that incremental goals lead to success at larger goals. Solution-Focused Brief Therapy's really all about goal setting to a large extent.

In Solution Focused Brief Therapy, therapists show clients how to view their problems in a different way. Your depressed client has never thought about why it is that they still have a job. A lot of depressed clients don't get out of bed, don't go to work, but some have maintained their employment for ten, fifteen, or twenty years. So, what gives them resiliency to go to work everyday even though they have depression? That's something they've never thought about before, but in thinking about things from a different perspective, our clients discover solutions that are truly useful.

The final assumption in Solution-Focused Brief Therapy's the clients want to change and have the skills to make changes. It's really important. I meet a lot of therapists each year, and quite a few have to some extent become cynical about the work. This presupposition is very encouraging.

Clients come to us in therapy ultimately because they want to make changes. And people actually do have the skills to make changes. The research shows that. They might not have the same resources all other clients have or that you have. They might not solve problems in the same way we do, but for each one of our clients, there is a solution to their problem deep within them. It is usually expressed in actions that they have previously taken in other situations and scenarios.

In Solution-Focused Brief Therapy, the role of the therapist is really kind of unique.[35] Along with the belief that people are resilient and have the strength and resources to change, therapists adopt an attitude that is respectful, positive, and hopeful. The therapist helps clients look for times when a problem wasn't present, and then asks, "Why wasn't the problem present? What were you doing then that was working for you?"

---

35 Steve De Shazer, Yvonne M. Dolan, and Harry Korman, *More Than Miracles : The State of the Art of Solution-Focused Brief Therapy*, Haworth Brief Therapy Series (New York: Haworth Press, 2007).

In Solution Focused Brief Therapy, we engage in exception-seeking, looking for times when a problem could have occurred, but didn't. In my relapse prevention work with alcoholics and drug addicts, I've often said to my clients, "Is there a time in the past year or two when normally you would have gotten drunk, or you would have gotten high, but for some reason you didn't?" And for almost every one of the addicts I've worked with, I've said, "Even though you might drink on a daily basis, there probably was some period of time for whatever reason when you did maintain sobriety for a couple of days, even unintentionally." And so I can find in that situation or scenario what resources were working for them. Why didn't the problem occur even though it normally is occurring?

In Solution-Focused Brief Therapy, even though we do a lot of questioning, Therapists rarely interpret the answers. There is not a lot of confrontation or challenge in Solution Focused Brief Therapy. Again, I am trying to help our clients become the best *them* they could be, rather than the best *me* that they can be.

Solution-Focused Brief Therapy is future-focused, and it compliments the client. It encourages clients to continue doing what is already working for them highlighting their strengths.

Stages of Change

One of the things that I really appreciate about Solution-Focused Brief Therapy is that it is stage-oriented. In other words, it looks at the counseling process with a beginning, a middle, and an end. As therapists, we ask ourselves what stage is the client in because at different stages of experience, our clients have a different set of solutions that are useful to them.

During therapy, clients may move in and out of what Solution-Focused Brief Therapy refers to as stages of change. It's important to recognize that these are guides, that clients can move forward and backwards in the stages. They can even move forward in some areas and stay the same in other areas. But I like having a model of change that sets out stages because when I find myself at the stuck point in therapy, it's at this point that I ask myself what stage my client is in. That question yields for me the information on what interventions I should be using with the client.

The first stage is pre-contemplation. At this stage, a client has no intention to change or to take any action towards the near future. They're unaware of the consequences of their behavior and avoid talking about their behavior. They're likely to underestimate the advantages of changing and overestimate the cost. These clients are not usually in therapy unless it is mandated by a court or strongly encouraged by family members. Most people in this stage are engaged in a process of change before they enter into our office for the most part. It's during this stage your client may have tried a variety of different solutions and experienced various levels of success. So when they finally make an appointment and come to your office, you can review what most likely has been the pre-contemplation process in the weeks or months that preceded their visit. This is a great place to explore solutions.

Then we have the contemplation process. This is where clients intend to change in the next couple of months. They're somewhat more aware of the advantages of changing. They're more accurately informed about the cons. They might be ambivalent about their situation or not quite ready for action-oriented treatment. Clients sometimes come to us in the contemplation stage because they're looking for information and exploring resources. They're really trying to prepare for a struggle that they perceive they might have. One of the most effective interventions here is to let our clients know that they've already found the solution and that by coming to your office, they've actually already embarked on the process of change. At that point, asking them the question, "Wasn't that easy?" can be a truly powerful tool for change.

The preparation stage is the stage where clients are really planning to take action in a concrete way in the next couple of days, weeks, or months. They are where they need to be to change, and they're ready to act on it. This is a great time to be goal-setting in Solution-Focused Brief Therapy. It is where the majority of clients end up coming to us in therapy for the first time.

Then we have the action stage: Clients have taken specific actions to modify their lifestyles within the preceding couple of weeks or months, so they are now in the process of making changes. This stage is when most couples enter therapy.

Through a series of actions, strategies and action steps, changes are implemented, so we find ourselves in the maintenance phase, in which clients work to maintain changes in their life and to prevent relapse. They have developed strategies to continue their changed lifestyle, and they're feeling more confident to continue.

Every therapeutic process has a termination stage. Ideally, our goal was to get rid of our clients. And so in this stage, clients are at the stage where they have little temptation to revert back to previous behaviors. They're confident. They're self efficient and they're ready to move forward. In the termination phase of therapy, we want to make sure that our clients realize that while they can continue to regard us as a resource, they should be encouraged to employ solutions apart from therapy.

Clients move to these stages at various times in their life with various issues. It's common for clients to move back and forth; however, by evaluating or assessing what stage of change our client is in, we can gain valuable information about what questions to ask and what processes to enter into.

Goal-Setting

What is a goal? A goal is concrete. It's clear. It's specific. It's time-oriented. It is beneficial. Most therapeutic approaches aim to develop clear, specific, and achievable goals for the client. In Solution-Focused Brief Therapy, the therapist attempts to make small goals rather large ones. The clients are encouraged to frame goals in a solution-focused way.

I do not write goals for my clients, by the way. In fact, I quit writing treatment plans I think in the early 1990s. Instead, I have my clients write their own treatment plans. I teach them how to goal-set, what goal-setting is all about, what objectives and goals are. I teach them that goals are always written. If they're not written down then it's just an idea or a dream. That's where our goals come from, but goals are written. If I have a client who in talk therapy establishes a goal, I reach over to my desk, I grab an index card and actually have them write that goal down.

I tell my clients to take that goal to tape it to their bathroom mirror. The reason why again, it's the first thing they see in the morning, the last thing they see at the end of the day. In fact, I've done this with many clients. I've told them, "I want you to get in the habit of writing goals and so I want you to take a stack of index cards and write a goal for each day, each and every day, for the next thirty days. Then just tape each goal to your bathroom mirror and don't take them down."

Now, it does not matter if my client never reaches any of those goals. They will in fact reach some of them, and others will be forgotten the minute they walk out the door in the morning. The purpose of this assignment is not so that they reach all of the goals that they set. It is so they get in the habit of being a goal-setting person.

I think that this is probably one of the most powerful strategies in interventions that I use in my therapeutic processes. In fact, if you take nothing out of this book, take that assignment with you because it is a powerful tool for change. By the way, it works wonderfully in couples counseling as well.

Identify Your Client's Strengths

An important role of the therapist in the Solution-Focused Brief Therapy model is to identify the client's strengths and resources. I actually created a tool called the Nongard Strength and Resources Inventory, which is available on my website. It is one of the forms I give all of my clients during the intake process because it tells me what they perceive as their strengths, what they perceive as their resources and that's my starting point in that very first session with the client for helping them make change

Strengths are of course those internal attributes that help a person to be successful. For example, are they trustworthy, loyal, helpful, friendly, courteous, kind, obedient, cheerful, thrifty, brave, clean, reverent? Resources are the specific things that are available to me to solve the problem. For example, if I am unemployed and looking for a job, and I have a cell phone, that means I can actually take the call when somebody calls me for an interview. I have a car so I can get to an interview. I have a resume that actually has advanced degrees on it. I have a professional license that allows me to work.

Clients might not have the same resources that I have or that you have, but I've never met a client yet with no resources at all. Identifying those resources is part of effective Solution-Focused Brief Therapy.

We can also take a questioning approach. Ask your client, "What are you good at? What would your wife, father, child, or other person close to you say that you're good at?"

Clients might actually be stumped by those questions. In fact, in my intake form, it asks what are your three strengths and a lot of my clients actually leave that blank.

During a session, the first thing I do is flip over to that page and see if they actually filled it out. If they didn't, I let them know this is the most important question on the form, and I press them for an answer. Sometimes, I have clients whom I see doing the intake form, and they're just sitting there. They have written down one or two things down. They're trying to come up with a second or third thing.

My clients have often never thought about what they're good at. They only think about what's wrong. In fact, I don't believe that I can fix most of the problems in my clients' lives. Their problems are just way too deep and way too difficult, but what I can do is help them use the resources that exist in their life to compensate for those deficits.

The Miracle Question

Solution-Focused Brief Therapy is probably best known for the Miracle Question. The Miracle Question is probably the leading technique in Solution-Focused Brief Therapy. As some clients have difficulties articulating a goal, the Miracle Question is a way to ask for a goal that the client comes up with by considering a preferred future.

The Miracle Question is a technique that helps clients to think broadly about new possibilities for the future and to imagine how their life would be changed if the problem was solved.[36] I also think the Miracle Question can be used as a tool for measuring success, but we'll come back to that in a moment.

Here is the question:

> Now, I want to ask you a strange question: Suppose that while you're sleeping tonight, and the entire house is quiet, a miracle happens. The miracle is that the problem that prompted you to talk to me today is solved. It just goes away. However, because you're sleeping, you don't know that the miracle has occurred. So when you wake up tomorrow morning, what will be different that will tell you that a miracle has happened and that your problem has been solved?

---

36 Linda Metcalf, *The Miracle Question : Answer It and Change Your Life*, 1st Ed. ed. (Williston, VT: Crown House Pub., 2007).

That really is a pretty complex question, but it's a great focusing question. The premise of the question is, of course, that a miracle happens and because you're asleep you don't know that it happens, so when you wake up, you're going to have to observe, feel, note, or think something different in order to know that a miracle occurs. What is that? What's different? And this can help us to establish a goal.

Likewise, the Miracle Question can also help us to define success because a lot of clients don't realize when they have made tremendous progress. They forgot the Miracle Question I asked in that first or second session. Now, I'm in the fifth or sixth session they're feeling stuck. But of course, I remember what their answer was, so I ask them about those things that I now know are in fact different. That way, I can help my client to accept where they are today and the progress that has been made even though they have not yet attained perfection.

There are follow up questions that we can ask after the Miracle Question, and these are really descriptions of preferred future needs. They can be broken down into smaller details and then translated into smaller goals.

Here are some examples of questions that could be asked following the Miracle Question:

- How will that be different?
- What will be different to you?
- What will you be doing instead when you are not doing _____?
- When you stop doing _____, what will you be doing then?
- When you are feeling _____ (the antithesis of the problem), what will you be doing? [That's a great question to connect an action to a positive emotion.]
- If someone were making a documentary film about you, how will she or he notice what you are feeling?

- Who else will notice you being more _____?
- What will they do when you _____ (stop smoking, quit drinking, stop chasing cars, whatever it is that they changed)?
- What will you do when she or he _____?
- What would be the first sign that he or she _____?

Any of these questions can be followed with "What else?" So that our client can really generate further alternatives at a deeper level. To some extent, this is very similar to the approach of Virginia Satir in family therapy.[37] Unlike Milton Erickson, who asked the broadest, most general questions possible, Satir would ask the most detailed and specific questions possible.

Progress Questions

In a previous chapter, I wrote about scaling questions—using a Likert scale to ask clients to rate their position on a scale of 1 to 10. That is certainly a part of Solution-Focused Brief Therapy as well. Asking our clients progress questions in subsequent sessions is a great idea to use.

For progress questions, the therapist asks the client again where they are in a scale to see if there has been progress, and then asks questions such as, "What's different since the last time we met? What has changed since our last meeting?"

Notice this is different than the typical introductory therapy question. Often the first question of therapy is, "So what would you like to work on today?" or "So tell me what's happened since our last session?" Asking what's happened since the last session is really living in the past, but asking the specific question about what is different since the last meeting

---

37 Virginia Satir, *The Satir Model : Family Therapy and Beyond* (Palo Alto, Calif.: Science and Behavior Books, 1991).

is a comparison that stays in the present. Some practitioners even begin each session with "What's better than it was last time we met?"

By using scaling and comparison questions, the therapist can then get a detailed description of what's better and how clients were able to implement the changes. Even if they didn't move from a level two of success to a level ten of success, we can use that scaling and that small change to ask questions looking for even more solutions:

- Now, that you're at a four, how are things different?
- What is it that you're doing differently?
- Who else might have noticed you being at a level four?
- How did it happen that you went from a three to a four on the scale in just one week?
- How did you decide to do that?
- How do you know you can do more of that?
- What needs to happen so you can do more of it?
- As you continue to do good things for yourself, what difference will that make to you from tomorrow?

These all provide opportunities for the therapist to compliment the client on being able to make things better and solidify change.

Exception Questions

Exception questions aim to empower clients to find solutions for their problems. Through the use of specific questioning, the therapist can help the client to identify times when things have been different for them. Exception questions often flow from the Miracle Question once a detailed picture of the preferred future has been attained.

It's important in the role of the therapist you are continually screening the client for talk about previous problem-solving in exception behavior.

This of course requires attentive listening to skillfully identify the client's previous solution behavior.

Here are some other great exception questions in Solution Focused Brief Therapy:

- Tell me about the times when you haven't been depressed.

- When was the last time you feel you were coping better?

- Was there ever a time where you and your partner communicated well?

- Can you think of a time when this problem was not in your life?

It is important that the therapist get details about the exceptions to help the client explore how they managed to be without that problem in the past.

Therapists are often struck by how easy it is to begin incorporating Solution-Focused Brief Therapy into their sessions. An easy way is to actually print some of the questions out on a note card and just set them on your desk so you'll easily be able to glance at them.

CHAPTER EIGHT

Mindfulness-Based Stress Reduction (MBSR) And Other Mindfulness-Based Therapies

Positive Psychology certainly is predicated on Mindfulness, as is ACT therapy, Dialectical Behavioral Therapy, and additional strategies addressed in this book. But I often hear people talk about Mindfulness-Based Stress Reduction (MBSR) as if it were a therapy.

However, Mindfulness-Based Stress Reduction is not therapy. It is the specific name of an eight-week training program that's advocated by the University of Massachusetts Medical Center as it was developed by Dr. Jon Kabat-Zinn at the University of Massachusetts. Its efficacy—its impact on those who graduate from the eight-week training program—has been meticulously documented. MBSR can also refer to an intensive five-day inpatient training module offered by the University of Massachusetts. This model of an eight-week outpatient training program or an intensive five-day inpatient training program has been adapted as a resource in pain control programs throughout the country, in almost every oncology center throughout the country, and in a wide variety of psychiatric settings.[38]

A Ph.D. neuroscientist, Jon Kabat-Zinn employs an approach of training and teaching clients rather than a counseling them. He began his work by treating those in the chronic pain program at Massachusetts General Hospital. The success in that particular area led to empirical studies of its utility with a wide variety of clients in varying contexts.

What is Mindfulness then? Well, Mindfulness as we discussed is the art and practice of paying attention to this moment. There's a famous saying that really captures that idea well. It's credited to Eleanor Roosevelt, but many people heard it recently from the great Master Oogway in the animated film, Kung Fu Panda:

> *Yesterday is history.*
>
> *Tomorrow is a mystery.*
>
> *Today is a gift,*
>
> *and that's why we call it the present.*

Mindfulness is really about staying in the present and experiencing the gift of this moment.

Mindfulness meditation may focus on anything, but we often focus on the breath, because the breath is always with us. Also, as long as we are breathing, something is going right.

Every time I work with somebody who has a fly-o-phobia, a fear of flying, I teach them Mindfulness because as long as they are breathing they are okay. In reality, as long as we are breathing right now, no matter what else is going on around us, we're actually okay.

In this exact moment, we're whole, complete, and safe. Mindfulness in clinical practice helps the client detach from fear, trauma, impul-

---

38 "What Is Mindfulness-Based Stress Reduction?" http://www.mindfullivingprograms.com/whatMBSR.php (accessed March 8 2014).

siveness, and self-defeating thoughts. It's a strategy that *trains* a client is in a new way of experiencing life, thoughts, emotions, and physical sensations.

There is a ton of evidence to back the efficacy of Mindfulness as either a primary treatment strategy or as a complementary strategy in any form of psychotherapy and medical treatment. There's a great deal of research as a matter of fact into both religious and secular forms of Mindfulness Meditation, but research in the clinical setting has shown that Mindfulness practice actually improves the immune system. It alters activation symmetries in the prefrontal cortex, a change previously associated with increase in positive affect and faster recovery from negative experience. Essentially, Mindfulness activates resiliency. This is why the U.S. Military has become very interested in preventing PTSD through resiliency training using Mindfulness techniques. A University of North Carolina at Chapel Hill study demonstrated correlation between Mindfulness practice in couples and enhanced relationship qualities.[39] Mindfulness-Based Stress Reduction programs have decreased relapse into depressive episodes by over 30 percent.[40]

The core concepts of Mindfulness are certainly a part of ACT therapy, Mindfulness-Based Cognitive Therapy, Dialectical Behavioral Therapy, and Contextual Hypnotherapy. Mindfulness provides a set of training techniques and strategies that integrate well into virtually any treatment approach.

The purpose of Mindfulness is to put space between our clients' thoughts, their emotions, or their sensations. Chronic pain clients become their pain, so Mindfulness as a strategy helps them to put some space between them and their pain. As Steven Hayes says, pain is unavoidable. Putting some space between them and their pain might not

---

39 James Carson et al., "Mindfulness-Based Relationship Enhancement", Association for Advancement of Behorior Therapy http://www.bemindful.org/mbrelenhanc.pdf (accessed March 8 2014).
40 J. D. Teasdale et al., "Prevention of Relapse/Recurrence in Major Depression by Mindfulness-Based Cognitive Therapy," *J Consult Clin Psychol* 68, no. 4 (2000).

stop the pain, but it can make all the difference in the world. Our depressed clients need some space between them and their depression. They become fused to that depression.

Because the word meditation is often used with Mindfulness, many people misunderstand Mindfulness. People think that meditation is about you know, sitting in a funny position, holding your arms out, saying a mantra, and trying to clear the mind. Mindfulness meditation is different from mantra meditation or transcendental meditation, which aim to clear the mind. The purpose of Mindfulness Meditation is not to clear the mind, and it is not to stop thinking. It's to practice simply noting when you do think and returning the mind back to the present.

Meditation is the process that is most often used to teach Mindfulness. However, meditation is not the goal. People are not better Mindfulness meditators because they spend thirty, forty, sixty, or even one hundred minutes a day in meditation. It doesn't matter if the meditation is thirty seconds or two minutes, or if it's done walking, sitting, or eating—because the meditation time or setting is not the goal. The goal is to learn a strategy. The goal or the purpose of Mindfulness Meditation is to develop certain traits that then can be applied in other contexts. The point of Mindfulness Meditation is to increase clarity and equanimity—evenness of temper even under stress. Mindfulness practice literally trains the nervous system in self-regulation. That's why there are so many tremendous benefits for our clients with physical difficulties. It's almost like a form of mental biofeedback, training the nervous system in self-regulation.

The purpose of Mindfulness is not only to ameliorate physical or mental symptoms, but to produce fulfillment in physical and mental processes. It's not simply to make what is difficult easier to accept, but it's actually to make life more wonderful.

My Personal Mindfulness Experience

Now, I'm going to share with you a personal story of how I actually developed a state of Mindfulness and understood what it was after many years of practice on the South China Sea.

I've been teaching and practicing Mindfulness Meditation for years yet it wasn't until 2008 that became personal for me. See, I found myself on a charter boat ride in the South China Sea, which was a perk for staying as a guest at a resort hotel. The vessel was a tiny sailboat. It was operated by a life-long fisherman who earned extra money giving rides to hotel guests.

Before we got on this boat, I thought to myself, "Hey those clouds don't look so great." Having lived in tornado country all my life, I understand what dangerous clouds look like, but he assured me everything was fine. Trusting that he was a professional and an expert in local weather, I got in the boat against my better judgment.

Before long, the sky grew darker, and the seas got rougher. It soon became clear that he had lost control of the boat, and we were in danger. As each wave doused us with powerful blasts of water and the rain began to hit my face, I actually started to panic. I knew I was way too far out to swim back to land, and the seas were so rough, I surely would have been swallowed up anyway if I jumped overboard. When I looked at the boat captain, this man of the sea, and I saw fear in his eyes, I knew death was soon to arrive. We drifted further and further from land and as control over the boat became nearly impossible, I could feel my body actually shaking in panic. In my own mind I saw anguished images of my children, mourning my death and asking why I was on the darn boat right in the first place.

I felt sad. I was fearful. I looked backwards and forwards wondering if I should jump. I looked to the sky, and suddenly I got pelted by this huge raindrop. It was a giant, and it was hot. It was actually a big hot raindrop. It was tropical rain, so it was salty and hot. With the boat moving, each raindrop splashed with a full force, the heat stinging my face. And it was in that moment that I really felt the drops, the salt, the heat, and the force of the rain, and it truly hurt.

The sea was white below my seat. It was a mesh seat, and the water was splashing underneath me as the wind blew ferociously. As I felt those things, I realized in my thoughts that I'm going to die. And I thought, "I don't want to die on this boat. I'm not ready to die today." The thought of my kids continued to pass through my mind. I could even see them standing with frowns and tears in my memorial service, and I was really suffering anguish on this boat with death probably just a few minutes away—and then I got hit by another raindrop, hot and salty.

And again, I noticed it and I said, "That was hot." In noticing that, I recognized at that moment that what I had always told my clients was true. No matter what else is going on as long as you're breathing, this moment is okay. So I breathed, and unintentionally I did what I have asked my clients to do in difficult situations: I took another breath. I didn't try to speed up or slow down the breath. I just breathed. And again, I breathed, paying attention to the wet, salty breath.

I noticed my fear, as well as my powerlessness over my fear, and I learned what acceptance was, simply accepting my fear as fear. I took another breath, breathing in, breathing out, fearful but breathing, breathing each breath into each moment that was left. Soon I noticed the feeling of the waves below my seat, the feeling of the wind on my face, and in that moment I was amazed by the beauty of the rough churning seas despite the impending doom that they would bring. I breathed again, and I felt the hot rain.

I hardly noticed when I stopped following my thoughts of panic, and I just experienced each moment not knowing which one would be the last. Then I noticed that my heart rate had slowed. Although my panic was there it became unimportant because I paid no attention to it. I accepted death simply noting my awareness of mortality, and in that moment I felt human. I felt a part of the sea. And I just breathed again, and I understood Mindfulness experientially, connecting each moment in just being okay.

I still have no idea how the boat captain got the boat under control, but somehow we made it to a small little island dot, and I was safe. We stayed there for a long period of time on the shore. We just sat in the rain. I breathed again, paying attention to the breath that one single moment—and I smiled.

And this is what will happen to your clients when you teach them Mindfulness in your office. There will come a point where what you have been teaching them becomes something they own. Every time I teach Mindfulness to clients, none of them says, "Wow, that was earth shattering and dramatic: that was just crazily life-changing." They almost all say, "That was okay. I can see how that can be helpful." Or they might say, "Yeah, I found that relaxing." But after practicing Mindfulness each and every day, there will come a point where they'll recognize and realize that they have been intuitively mindful. It's at that moment that they will own Mindfulness—and they will own Mindfulness from that point forward as well.

Day-to-Day Mindfulness

Let me give you a couple of thoughts about Mindfulness in daily tasks:

- You don't have to strive to overcome anything. Mindfulness is about helping a person to overcome things, not through striving to do it, but by really just being fully present. The result of Mindfulness is that overcoming just happens. It's a natural outcome of being present in each moment.

- Mindfulness can be done in a number of different ways. Paying attention to daily activity, you can mindfully brush your teeth, mindfully eat a raisin, mindfully walk, or mindfully sit and breathe.

- Mindfulness really is the practice of redirecting our attention. It is learning a technique for being able to say, "That's a thought," or "That's emotion," or "That's a sensation." Instead of simply following that experience mindlessly, you can recognize that that thought, feeling, or emotion is present, and then use it as an indicator to bring your attention back to the present.

- Mindfulness Meditation is ultimately about learning how to note spontaneous and intuitive Mindfulness. There's an old Zen teaching that says the most important thing is remembering the most important thing. And that really is the essence of Mindfulness.

Singing in the RAIN

Now, the mindful presence that helps release emotional suffering is summarized with an acronym in Mindfulness-Based Stress Reduction. I think it's real appropriate based on my story in the South China Sea. and that acronym is RAIN:

- Recognize when a strong thought, emotion, or sensation is present. When I teach clients how to practice Mindfulness, I'm teaching them how to note their thoughts and feelings non-judgmentally. The reason our clients become fused or enmeshed with their thoughts is that they judge. They assign meaning: I felt angry, and that's bad or I felt angry, that means this or I felt angry, and this is what happens when there's anger. Those are relational frames. The relational frames are really our mind following a thought, feeling, or emotion. Non-judgment in Mindfulness-Based Stress Reduction is about training our client to simply note a feeling, note a thought, note a sensation, then return their thoughts to the present.

◆ Allow, acknowledge, or accept that the thought, emotion, or sensation is there. So much of what happens in our life is really from the subconscious mind and even the unconscious mind and practice recognizing what is arising, fear, hurt, sensations, thoughts and then allowing those things to be present, to be with it, to be in the same room as hurt or anger or fear to allow it to be present. The I stands for investigate in a non-analytic way of getting to know the body, the heart, the mind and experience those energies.

◆ Investigate the thoughts, emotions, or sensations. When I work with pain control clients and teach them Mindfulness, investigation is very important. I want them to investigate their body in the present moment, and they'll almost always discover something. They'll almost always discover that when pain is present, so is health somewhere else in the body.

◆ Non-identify with the thought, emotion, or sensation. The point of non-identification is that you're not having your sense of being defined by, possessed by, or linked to any emotion. In other words, you're not taking it personally. Think about how many of your clients particularly in couples counseling need to learn this.

Characteristics of Mindfulness

In his book *Full Catastrophe Living,* Jon Kabat-Zinn gives us eight qualities or characteristics of Mindfulness:

1. Non-judgment: Simply allowing a thought without having to believe something about that thought.

2. Patience: Appreciating this moment rather than longing for a future moment.

3. Beginner's mind, curiosity: Approaching the familiar with a sense of curiosity. (That's why I love teaching clients Mindfulness based on the breath. You know we've been breathing since the first day of life and we breathe until the end of life, yet we never think about breath.)

4. Trust: Honoring your feelings, intuition and trusting yourself.

5. Non-striving: We're not really trying to get anywhere with Mindfulness. We're just trying to be mindful.

6. Allowing: Letting the present be experienced without an agenda or goal.

7. Acceptance: Seeing things as they really are in the present.

8. Letting go: It's how we go to sleep at night. It's how many people go to work—at least those who do work that they don't particularly enjoy. If we go back to Solution-Focused Brief Therapy, letting go is probably a skill most of us already have. Being able to apply it in the context of other situations can be a particularly useful strategy.

I don't know where I heard this, but somebody said that when we're old and in our hospital bed looking back on life, there will be one hundred days that defined us, the days that made us who we were. Most people will not recognize those hundred days until they look back on their life in retrospect. Mindfulness is all about teaching a person to recognize those hundred days when those days are happening.

Mindfulness in Action

Let me give you an example of how Mindfulness can be useful. A few years back I was driving from Tulsa to Wichita with my son. It's a three-hour drive, and there's not much between Tulsa, Oklahoma, and Wichita, Kansas. But at the halfway mark there is a gas station and convenience store on the turnpike, and of course there is a McDonald's there. Now, my son and I usually try to eat healthy food, and we were looking forward to eating at a delicious Thai vegetarian restaurant in Wichita. However, we had left in the late afternoon, so we were certainly nearing the dinner hour when we were still a hundred miles away from Wichita.

When I stopped to buy gas, my son said to me, "Hey, Dad, I'm hungry, so I'm going to go in and get a burger from McDonald's real quick."

And I said, "Okay, if you want to be judgmental."

"What?" he said.

I reiterated, "I was just pointing out that you were being judgmental."

He said, "What do you mean being judgmental?"

I said, "Well, you said you were hungry."

"Right, I am," he answered.

I said, "Then you said you were getting a cheeseburger, right?"

He said, "Yes."

"Well, you know we're going to be eating at the delicious Thai vegetarian restaurant in Wichita in another hour and a half or two hours."

He said, "Yeah. I'm just going to get something that tide me over."

"Okay, you know, I was just pointing out that you're judgmental," I commented.

And he said, "What kind of therapy stuff is this, Dad? I'm not being judgmental."

So I said, "Well, you had a sensation, right? Hunger?"

"Yeah, I'm hungry."

"So, when you had that sensation," I said, "instead of just being hungry, you judged the hunger and you came to the conclusion that you must solve it now. You came to the judgment that hunger is bad, so you need to eat something immediately so that this feeling goes away. That's judgment, and I just wanted you to know that that's why you're getting a cheeseburger. It's because you've been judging your sensations."

He kind of looked at me with this puzzled and yet semi-enlightened look as he said, "What am I supposed to do about it?"

I replied, "Well, you can do whatever you want to do about it."

He said, "Well, I want a cheeseburger."

I said, "Okay. Well, then go grab a cheeseburger and be judgmental."

"So, what else am I supposed to do?"

And I said to him, "Well, here's what you could do: You could just be hungry. You could observe hunger and instead of attaching meaning to it or judging it and taking an action based on the feeling, you could just see it as it is: hunger, which is simply a natural part of the digestive process."

He said, "How am I supposed to do that?"

And I said, "Here, go ahead, close your eyes. Now, take in a breath; breathe in and out. Pay attention to the breath, and now notice that place in your body where you're feeling hunger. Really become mindful of the sensation of hunger. Observe it. Breathe in and breathe out. With each breath, really just focus on that feeling of hunger in the body.

"Now when you notice that feeling or sensation, rather than following it and saying to yourself, *I need a cheeseburger,* or *this is bad,* or *this is intolerable,* instead just say, *I note that sensation. That is hunger.* Just call it by its name and say, *that is hunger, that is hunger* and just breathe and be hungry without being judgmental. Say to yourself even *this is what body does. It digests. This is how digestion feels.* Call it by its name. *This is hunger.*

"Breathe in, breathe out, and simply note what hunger feels like. If you notice your mind wondering to the thoughts of a McDonald's or a cheeseburger just note that thought–you don't have to judge it. You don't have to say it was bad. Say *that was a thought* and bring your attention back to your breath or to that place in your body where you feel hunger. Even if you have to do that whole bunch over the few minutes, that's okay. Just be fully present, and note that in the present moment, you're hungry."

Of course, we were standing in the doorway of a gas station and probably looked a little bit silly, but when he opened his eyes, he had a big smile on his face. His affect had changed. He looked fantastic. I walked towards the car. I noticed he was walking right behind me. He got in. I started the car up, and we drove off.

We got onto the turnpike, and he said to me "Hey, you tricked me."

I smiled and laughed. I said, "No, I helped you understand Mindfulness."

So you can learn to experience Mindfulness in a number of ways. You don't have to go on a boat to the South China Sea. You can actually go to the middle of nowhere, Oklahoma, stop for gas, and learn these things. Mindfulness is a skill that is taught.

I've worked with a lot of eating disorder clients over the years who actually need to learn that hunger is just a part of digestion. It's not something they need to interpret or take action on or get rid of. When my clients with eating disorders or my obese clients trying to lose weight learn that, they experienced success at weight loss no matter what kind of diet they're on.

Mindfulness can be practiced in many ways. We can do a walking Mindfulness Meditation. I often take a walk, and I'm aware of my feet and I'm aware of my walk and the temperature. It's great even though I walk around the same neighborhood every night. In the neighborhood where I live I almost always discover something new when I'm mindful.

I use the Raisin Meditation with my clients because I treat a lot of obese clients, and so this is one way that I teach my clients some really wonderful things about Mindfulness and about their eating.

I learned about the utility of Mindfulness for weight control back in the early 1990s when I was a part-time family therapist in the William Rader Eating Disorder Treatment Program. Up to that point, I had never thought about my own eating. I was number five out of seven in a large family, so I learned to eat fast because if I didn't snag as much pizza as I could as fast as I could, I probably wasn't going to get a second or third piece.

So, when I started working as family therapist, I was expected not only to conduct group, but to also eat dinner with them. Dinner was scheduled from 5:30 to 6:30—an entire hour. The first day I was there I actually chowed down my meal in about three minutes like I was used to. All the clients just sort of stared at me. One of the other therapists explained that dinner time was in fact therapy time, and that we were going to learn that the purpose of eating dinner together as a group was to learn to pay attention to our eating and to eat slowly so as to change those unhealthy habits. I learned the value of eating mindfully way back then, and I've worked to internalize those concepts in my own life.

The Raisin Meditation

If you don't have a raisin, you can do this with a small piece of chocolate. Dark chocolate is of course healthier. You could do this with an M&M. You could do this with really a small piece of fruit or anything.

> *Hold a raisin in your hand and really bring your attention to that raisin. Chances are pretty good in the past you've just grabbed the raisin and chowed it down. You've probably done that with a whole handful of raisins. In fact, you've probably never eaten just one raisin before. But observe that single raisin as if you'd never seen one before. Touch the raisin with a finger. Is it rough or smooth, thick or thin, hard or soft?*
>
> *Feel the raisin, the texture between your fingers. Feel the weight of the raisin and even notice with your eyes the colors of the*

raisin. Each raisin has multiple colors. Notice if you have any thoughts you might have about the raisin or any feelings of liking or disliking a raisin. Note the color of the topography, the scent of a raisin. Look into the valleys and peaks, the highlights and the dark crevices of the raisin. Now, lift the raisin under your nose and smell it. What does a raisin smell like?

Bring the raisin to your lips, being aware of the arm moving the hand to the position to correctly feed yourself a raisin. Mindfully be aware of the mouth salivating as the body is anticipating eating. Feel the emptiness of your mouth. Note the sensation of no sensation since you haven't eaten the raisin yet.

Now, take the raisin in your mouth and just allow it to sit there for a moment. Don't chew it. Move it slowly around with your tongue.

As the raisin lies in your mouth, notice whether the raisin is warm or cold. Notice the texture. Notice the rough edges, and notice the smooth surfaces of flavor. Notice, the subtleties of the flavor.

Now bite down once into the raisin. Hold it between your teeth. Notice the sweetness. Notice if you have a thought. Notice if you have an emotion. Notice a breath. Mindfulness is about noting things, becoming aware of things. And as you separate your teeth, what did you notice about the texture?

And as you bite down again, chewing again, notice how the texture changes as you chew. Are you aware of how warm the raisin has become? Has it become juicier? Notice any anxiety or stress you might feel by not swallowing the raisin yet.

And now slowly finish chewing and when the raisin is all but dissolved, go ahead and swallow and breathe and you can even say to yourself, "I have eaten a raisin."

I use this Raisin Meditation with my clients—not only in my bariatric clients, but really for many different clients to teach them the art of Mindfulness. Mindfulness is taught to clients by helping them to do something they have never done before: to become a full observer in the moment.

CHAPTER NINE

Dialectical Behavioral Therapy

Dialectical Behavioral Therapy (DBT) has been quite popular since the early 1990s, and like all the other approaches in this book, DBT is an evidenced-based treatment protocol. There is a mountain of evidence to show its efficacy in what it was first designed to treat: Borderline Personality Disorder.[41]

Most therapists and counselors have been extremely frustrated by personality disorder clients, almost viewing them as untreatable. However, that simply is not true. We can help our borderline clients move to a higher level of functioning by applying DBT in areas such as eating disorders, self-mutilation, and other fairly aggressive psychiatric conditions. DBT is a form of therapy that combines cognitive behavioral techniques and Mindfulness to teach clients to regulate their emotions, tolerate distress, and to improve relationships. Those are really the three chief goals.

I have always thought that one of the most important areas in psychology or counseling is to help clients learn how to tolerate distressing situations. Scott Peck made millions of dollars in the 1970s writing his book *The Road Less*

---

41 Marsha Linehan, *Skills Training Manual for Treating Borderline Personality Disorder*, Diagnosis and Treatment of Mental Disorders (New York: Guilford Press, 1993).

Traveled. It's not a particularly good book, but it made lots of money because it had a great opening line: *Life is difficult*. People opened that book and said, "Oh my gosh, this book is about me. I'd better spend the $10 and buy it."

Most people need to learn how to increase their distress tolerance skills, and DBT is a therapeutic approach that specifically does that. Dialectical Behavioral Therapy of course was developed by Marsha Linehan in Washington and published in 1993 as *Skills Training Manual for the Treatment of Borderline Personality Disorder*. By the way, the book was not called "Dialectical Behavioral Therapy" because the publisher didn't feel that would sell. Linehan recognized that the established Cognitive Behavioral Therapy structure had a limited effect on treating the core problems in personality's function. She observed that something was missing in CBT, particularly when dealing with individuals who engage in self-injurious behaviors and suicidal idea tendencies.

The additional techniques of Mindfulness which we've discussed in regards to many other therapeutic approaches teach people with borderline personality disorder to regulate their emotions more effectively, especially helping them deal with overwhelming negative emotions, which are characteristic of the borderline personality disorder client.

In contrast to Cognitive Behavioral Therapy, Dialectical Behavioral Therapy offers patients an extended therapeutic framework rather than a traditional brief therapy approach. Sometimes people say to me, "What about long term care?" and as I mentioned, there are some clients who have more catastrophic difficulties than others. Even though I tend to favor a brief therapy model, DBT is one of those models of Contextual Psychology that we can use on a long-term basis with our clients.

There's tremendous support for Dialectical Behavioral Therapy as a structured program. Like Mindfulness-Based Stress Reduction, DBT is a specific training protocol. It is a structured program that has been well researched and documented; thus, it is certainly considered an evidenced-based approached to therapy.

People who complete DBT programs have been found less likely to engage in deliberate self-harm and suicidal attempts over a year of observation. Participants in DBT had been found to have fewer inpatient psychiatric stays than those who didn't engage in the program, and this was also largely maintained in a one-year follow up.[42] That's really important for those who work with catastrophic mental illnesses. Individuals with Borderline Personality Disorder who underwent a protocol of Dialectical Behavioral Therapy had lower depression scores, less deliberate self-harm and improvements in anger, hostility, hopelessness, and dissociation.

Think about the clients you work with, if you could implement techniques that could reduce feelings of anger, hostility, and hopelessness, while ameliorating self-harm and decreasing depression, wouldn't you want to implement those strategies?

In this chapter I'm going to share the structure of Dialectical Behavioral Therapy and the stages of therapy, but I'm also going to address specific methods that can be useful to help you help your clients experience success.

Now, in a traditional program of Dialectical Behavioral Therapy there are two components and they're viewed as equally important: individual therapy and group therapy.

However, many professionals in private practice do not facilitate group therapy, and this could be for a variety of reasons. Their offices are too small. They don't feel like they can get enough clients together to form a group where it would be worth the time and commitment to the group for the financial rewards that may be available. Nonetheless, many of us do offer group therapy in the context of our private practice offices, and certainly in outpatient and hospital settings that is going to be one of the norms.

In Dialectical Behavioral Therapy, the individual therapy component

---

42   C. R. Swenson et al., "The Application of Dialectical Behavior Therapy for Patients with Borderline Personality Disorder on Inpatient Units," *Psychiatr Q* 72, no. 4 (2001).

involves the therapist and patient having an ongoing discussion about issues that come up during the week and recording it in a journal. That's really important. I'm a firm believer that when we write something down it becomes reality. I wrote about affirmations in a previous chapter. A thought needs to become words, and words need to become written.

Deliberate self-harm and suicidal behaviors are the first priority in individual therapy. One of the things I love about DBT is that it does not try to tackle every problem at once. It ranks the problems that clients have as part of the assessment process. Skills that are learned during group therapy are further reinforced during individual therapy, which is usually structured on a weekly basis during the first twelve months, then tapers off as people practice and implement these ideas on their own. It's important to remember that even when working with a client with a more catastrophic set of conditions or diagnosis, they're still ultimately our goal. Ultimately, we still aim to have them leave our caseload, to be able to function independently.

Group therapy is generally provided about two-and-a-half hours a week over a twelve-month program. Some groups might run bi-weekly. The goal of groups really is skill training. That's the main focus of group therapy with the underlying group dynamics and cohesion coming into play. In a lot of therapeutic modalities, underlying group dynamics and unanimity of the group is the goal. In DBT, it's really more of a patient education setting. Even if clients are at different functional levels, or enter or exit the group at different times, resulting in decreased cohesion, group therapy is still successful.

Stages of Therapy

DBT incorporates a model of treatment that identifies four key stages. Its approach prioritizes problems that are urgent and addresses these at various points in therapy. One of the core goals of DBT is to help clients create a life that is worthwhile to them at the same time as dealing with the task of gaining control over problem behaviors, especially those that are self-destructive.

Stage 1: Gaining Control of One's Behavior

The first stage of DBT is about moving from experiencing uncontrollable behavior to taking control of behavior. During the assessment process, a ranking of what behaviors are most important to address is made:

- Suicidal behaviors
- Behaviors that interfere with therapy
- Behaviors that impact on quality of life
- Behaviors to be addressed with skills training

For example, if your client has suicidal thoughts and actions, you deal with those first. Suppose your client is an alcoholic who gets drunk and comes to therapy—that is a behavior that interferes with therapy. If your client will not get out of bed and go to work they won't have enough money to pay their bills, which impacts quality of life.

Finally, stage one involves teaching my clients certain skill to help them develop awareness, improve relationships, understand their emotions, and tolerate emotional pain.

One of the things I love about DBT is the first question therapy is really a question for me. I know that a lot of therapists who say to their clients, "So what would you like to work on today?" or "What has happened since last week?" But really the first question is "What does my client not know that they need to know in order to choose to do something different?" It's my responsibility in the therapy session to teach my clients those skills.

These skills can be categorized like this:

+ Core Mindfulness Skills
+ Interpersonal Effectiveness Skills
+ Emotional Modulation Skills
+ Distress Tolerance Skills

I've worked in a lot of drug and alcohol settings, and I have met clients, believe it or not, who really don't know how to not drink. They don't know how to say no to others. I work with clients who, when they're depressed, really don't know how to do anything other than pull a sheet over their head and stay in bed. I have met clients who, when they find themselves in crisis, literally do not know any other way to call for help other than suicidal action. So I have to ask myself, "What does my client need to know in order to choose to do something different?"

At times, the answer to that question points to skills that may be very practical. For my client to change their life, maybe they need a job, so I may need to be teaching them how to get a job. There are clients who really don't know what they need to say on an interview, how to be assertive, how to kind of get their foot in the door and move from the receptionist to that person who can actually hire them. So skills training really is important.

I have balanced a lot of checkbooks with couples during couples counseling because they literally did not know how to manage money, and money became the therapeutic issue that was presented.

Stage 2: Experiencing Emotions Fully

Stage 2 of DBT is about helping my clients to experience emotions fully. The second stage aims to help clients experience negative emotions without reverting to dissociative, avoidant behavior or being overwhelmed by symptoms of trauma. The therapist works with the client in this stage to teach them how to experience all of their emotions without the emotion taking over. Techniques from a number of other contextual therapies could be particularly helpful at this stage:

- Mindfulness-Based Stress Reduction
- ACT Therapy
- Deliteralization
- Cognitive defusion

Cognitive defusion is especially important, helping clients to experience negative emotions without being overwhelmed and relapsing into impulsive and destructive behaviors. Those who work with PTSD clients find that during this stage, the client learns some basic self-control—something they felt had been taken away from them because of the trauma.

Stage 3: Managing Everyday Problems

In Stage 3, the goal of Dialectical Behavioral Therapy is moving towards managing everyday problems. Once they manage their emotions and have control over their destructive behaviors, clients can work towards managing ordinary everyday problems: those associated with relationships, work, study, finances. The client is given more space from therapy to try their skill out in the real world. This provides them with the space to synthesize the skills they have learned and develop healthy problem-solving strategies. Stage three aims to increase the client's self-respect and self-efficacy, thereby improving quality of life.

Stage 4: Being Connected

You know, nobody ever got well alone. An important focus in therapy needs to be moving our clients towards a state of connection with others, which is why Stage 4 of the DBT process is about being connected.

The final stage of DBT involves assisting a client to feel complete and connected to their world. Some clients might find that even though they've moved through stages 1, 2, and 3, and they might have their life to some extent the way they want it, they still feel empty or disconnected. So the fourth stage involves synthesizing the success of the former three stages into the client's identity. We can even draw from Solution-Focused Brief Therapy to do that.

Progress through the stages is not linear and will overlap at times during the therapeutic alliance. The purpose of DBT is to encourage the patient to acquire a set of skills to a sufficient level that they will have satisfactory quality of life and control over their behavior.

Mindfulness in DBT

We've talked about Mindfulness throughout this book—and for good reason. Within each of these different modalities of Contextual Psychology, there seems to be a foundational recognition that Mindfulness is a skill that can be taught to clients and a skill that they respond to effectively.

In DBT there are three primary states of mind that Mindfulness is used to address:

+ Reasonable Mind. This is when a client approaches things logically and engages in planning behavior and focused attention when approaching problems, for example planning for an event in advance.

+ Emotional Mind. A client is in the emotional state of mind when their thoughts and behaviors are heavily influenced by their feelings.

+ Wise Mind. When a client is in this state, their reasonable mind and emotional mind are integrated. They have intuition and a sense of what feels right and wrong. This comes from practicing

Mindfulness. It doesn't come from attending a lecture. It doesn't come from earning a psychology degree. It actually comes by practicing Mindfulness with intention on a daily basis.

Mindfulness Skills

Dialectical Behavioral Therapy teaches specific Mindfulness skills:

- Observation: The observant mind is the part of the mind that observes experience in the current moment. That's why we focus on the breath: It's something to observe.

- Description: The next part in learning Mindfulness is to describe the experiences that come into awareness so when the thought or emotion arises, the client learns to put words to it. I encourage my clients this way: When you have anger, simply say to yourself, "That was anger." If you have fear, say to yourself, "That was a fear." Thich Nhat Hanh teaches that the name of something is important, and so when we breathe in call it in; that's its name. And when we breathe out, call it out.

- Participation: Another Mindfulness skill to cultivate is participation. To develop mindful awareness, the client needs to practice the skills and so whether in group or individual therapy, participating fully is important.

- Non-judgment: Being able to focus on the what and not the interpretation of the sensation is the essence of non-judgment.

- One Mindfully: This is really a core Mindfulness skill that crosses the spectrum from Buddhism to psychology to every approach within the discipline of Contextual Psychology. It's a core Mindfulness skill that teaches clients to focus only on the present moment one thing at a time. By the way, things are not unique to one discipline and one discipline only. Some of you will recognize this from the Alcoholics Anonymous mantra, "one day at a time."

- Effectiveness: This aspect of Mindfulness teaches the client to focus on what works and what is really most effective.

CHAPTER TEN

DBT Skills Training

The approach of Dialectical Behavioral Therapy is to a large part predicated on skills training. That's why the first book[43] that Linehan wrote focuses on three things:

- Interpersonal Effectiveness
- Emotional Modulation
- Distress Tolerance

Interpersonal effectiveness is one of the most important skills to teach. This is because it can be used in a variety of contexts:

- Attending to Relationships: Effective interpersonal skills can be used to end damaging relationships, ask others for help, say no to situations, resolve conflicts, and address problems before they become overwhelming.
- Balancing Priorities vs. Demands: Learned skills help the client to prioritize more effectively, and reduce or defer things that are a lower priority. It also allows them to ask for help when needed and say no if they are feeling overwhelmed.

- Balancing the Wants and 'Shoulds': There are various things the client may want to do because it will instigate change or it is enjoyable. There are also things that they should do because it needs to be done or they feel it is expected of them. (This overlaps well with Albert Ellis's Cognitive Behavioral Therapy.)

- Building Mastery and Self-Respect: Skills will help the client to feel competent and effective when interacting rather than helpless or dependent. They will be assertive and heighten their self-respect, an important quality I think is often underrated.

I actually self-esteem and self-respect are probably the most effective skills that can be taught, and yet they are among the least often taught skills in psychology. These skills help our client feel competent and effective when interacting, rather than helpless or dependent. I love teaching my clients assertiveness.

In fact, back in the 1990s, I was training probation officers in a variety of subjects related to mental health, and I wrote a book called *The Getting Along Workbook*. It was a workbook to help clients learn to master the skill of interacting with others in a non-helpless and non-dependent way by using assertiveness. It is helpful in couples counseling, with criminal justice clients, with substance abusers, and with many other clients.

Interpersonal Skills

There are three main goals in relationships and interpersonal effectiveness skills training:

Objective Effectiveness

The goal of objective effectiveness is for the client to use interpersonal skills to obtain something that they want. Although it is not a guarantee that they will get what they want, it does teach clients to be assertive and resolve interpersonal conflict as well as have their opinion taken seriously.

Objective effectiveness skills are taught in DBT using the DEARMAN acronym.

- ◆ D Describe
- ◆ E Express
- ◆ A Assert
- ◆ R Reinforce
- ◆ M Mindful
- ◆ A Appear confident
- ◆ N Negotiate

Dialectical Behavioral Therapy, like some of our other approaches, seems to love acronyms. I have a dry-erase board in my office, so when I'm doing individual therapy with couples, I will often actually get up out of my chair and write on that dry-erase board: DEARMAN. That way I can teach them to communicate object effectiveness by describing, expressing, asserting, reinforcing, being mindful, appearing confident and negotiating. This is a great thing to teach the couples you work with at couples counseling.

Relationship Effectiveness

The goal of relationship effectiveness is to use skills effectively to maintain or improve a relationship. Clients learn that they can get what they want while maintaining a relationship with someone.

Relationship effectiveness involves:

- ◆ Behaving in a way that makes the other person want to comply with your requests.
- ◆ Behaving in a way that makes the other person feel good even though you have said no to their request.

- Balancing short-term goals in terms of what is best for the relationship. The goal may be to have the other person stop rejecting them or to approve of them; in these cases relationship effectiveness aims to do this in a way that improves (not damages) the relationship.

To balance short term goals with the longevity of the relationship requires interpersonal effectiveness. Attacking someone or being verbally abusive towards someone is a short term relationship gain that would risk the relationship in the long term. Relationship effectiveness skills can be taught using the GIVE acronym.

- G Give
- I Interested
- V Validate
- E Easy Manner

Self-Respect Effectiveness

The goal of self-respect effectiveness skills training is for the client to effectively maintain their self-respect. It aims to maintain and improve a client's positive feelings about themselves and respect their own values and their own beliefs. In the context of getting their needs met, they may learn to act in ways that fit with their morals and make them feel worthy. Self esteem effectiveness is also taught with an acronym, in this case, FAST:

- FAIR
- APOLOGIES
- STICK TO VALUES
- TRUTHFUL

The acronym DEARMAN can help clients improve their self-respect and sense of mastery as they learn to balance what they want with the other person's needs. GIVE skills enhance a client's self-respect, and these skills can be used at different times or in conjunction with each other. Our clients do need help prioritizing what is most important in each situation, whether it is getting what they want, maintaining the relationship, or having self respect.

Traditional approaches to therapy can help clarify those things, but teaching these skills to your clients will help them take better care of their relationships, take better care of themselves, balance their priorities, and build self-respect.

Emotional Modulation

Emotional modulation skills are very important in DBT. In order to modulate their emotions, clients have to learn how to understand what an emotion is, how it functions, and how to experience emotions without being overwhelmed. Overwhelm is one of the greatest frustrations our clients face. It's what precipitates crisis for many of them. Emotional regulation skills also help the clients recognize more clearly what they feel as they learn to observe each emotion without becoming overwhelmed.

Emotional regulation skills are designed to substitute destructive coping strategies with more effective strategies. Essentially what we're doing is we're replacing an old pattern with a new pattern. That's why the skills need to be practiced. A popular rule of thumb states that it takes twenty-one days to make a habit, which is why I often give clients the assignment to practice something on a daily basis for the next seven days—and then I'll have them continue that for the next seven days, and then for another seven days until we reach twenty-one days. It's my assumption that on the twenty-first day of practice they probably have incorporated the new skill into their lifestyle.

The goal of these skills is for clients to modulate their feelings without behaving in a reactive or destructive way. And the easiest way to stop a reactive or destructive behavior is not to try to stop it. That only creates a greater urge. This is the paradox of the yellow jeep exercise from Acceptance and Commitment Therapy. The only way to eliminate a reactive or destructive manner is to replace it and to have the replacement behavior practiced so that it becomes the new norm.

One of the main skills taught in emotional regulation is understanding emotions. We use words to describe emotions, yet these words really don't have a meaning to our clients. They have become their anger; or they have become their depression; or they have become their anxiety. For every adjective that we use to describe a feeling, the emotion could be experienced as either a primary emotion or as a secondary emotion:

- *I feel hurt, but that's because I'm scared* (the secondary emotion).
- *I'm angry, but that's because I'm hurt.*
- *I'm happy, that's because I'm connected.*

And so for every primary emotion there's always a secondary emotion and asking my clients to describe their emotions with an adjective and then to look for the accompanying emotions is a simple exercise, but a powerful exercise in helping my clients to see how these things are really coexisting and relating.

Another skill is teaching clients how emotions function. I explain that emotions exist for three reasons:

1. To communicate to other people
2. To motivate me to action
3. To be self-validating

The non-judgmental attitude of Mindfulness helps clients to avoid becoming overwhelmed. Emotions are neither good nor bad. They just are what they are. I let my clients have and own whatever emotion that they feel because that self-evaluation is important. I teach my clients how to experience emotions without feeling a sense of overwhelm. One of the ways to do that is to have your client practice Mindfulness to put some space between them and their emotion, to experience emotion as an observer rather than experience it from that first perceptual position.

Reducing emotional vulnerability is another key emotional modulation skill. Dialectical Behavioral Therapy tries real hard to make an acronym for this, and the acronym is PLEASE MASTER. It just doesn't make any sense to me; it's so hard to remember the acronym then you may as well just teach the attributes.

DBT tries to make an acronym out of the following:

- treat PhysicaL illness
- balance your Eating
- Avoid mood Altering drugs
- balance your Sleep
- get Exercise
- build MASTERy of emotions

Setting aside the tortured acrobatics DBT uses to make this acronym, what we're really talking about here is what Alcoholics Anonymous has been saying for years. In 12-step meetings, we have heard HALT—never get too Hungry, Angry, Lonely, or Tired:

- Hungry
- Angry
- Lonely
- Tired

That's an easy acronym to remember, but the concept here is the same. We are only as well emotionally as we are physically.

I got certified probably about ten or twelve years ago as a personal fitness trainer. Not because I was planning to work with clients in a gym or in some sort of fitness setting. In fact, up to that point in life, I really hadn't gone to the gym. I hadn't exercised. I hadn't worked out. But I recognized as I was getting a little bit older that I probably needed to go to the gym, so I joined the gym.

When I joined the gym, I got two free sessions with a personal trainer, and he was so helpful and valuable to me that I purchased more sessions with him. One day I was working out with him when I noticed that all the way across the gym there was a trainer standing next to a treadmill. On the treadmill was a person who was about four hundred pounds. They were just slowly walking on that treadmill, one painful step at a time, for about forty minutes.

Now, since I treat obesity in my office, I recognize that it is a clinical issue for psychology counseling and social work. However, on the drive home, I thought to myself, "Wow, that personal trainer actually just did a therapy session, but instead of talking to a client about the change they could take when they left the office, that trainer actually was participating in change with the client."

And I thought to myself (metaphorically, of course), "What if we replaced the therapy couch in our office with a treadmill? Even if I'm not treating obese clients, even though I'm treating depressed, addicted, or anxious clients, people only are emotionally able to function as well as they are physically." I became so intrigued by the idea I thought, "Forget about the metaphor; let's actually put a treadmill in the office!"

I never did put a treadmill in my office, but I actually became certified as a personal fitness trainer because I wanted that base of knowledge to help me make the connection between physical wellness and emotional wellness.

Another skill for emotional regulation is helping our client to build positive experiences. Part of the therapeutic process should be entering into positive experiences with our clients so that they have a basis for experiencing positive experiences in our office which translates into an ability to build positive experiences outside the office.

Another strategy for emotional modulation is the "opposite-to-emotion" action—an action that is completely the opposite of what the client is feeling. So, if a client feels sad, we encourage them to smile, laugh, and dance. The paradox here is that when we instruct clients to act on the opposite of the emotional urge they feel, it can help them to gain emotional regulation. Now, this is not something you were going to use with those who are high functioning clients who have a very few difficulties at this point or nearing the end of therapy. When we have clients in the beginning of therapy who are very impulsive, very quick to anger, who think, who act before they think, teaching them simply the paradox of acting the opposite of their urge is a great way to help them enter into the practice of modulating their emotions.

Of course, another technique is to help our clients check the facts. I tell my clients, "When you feel something, check the facts. Check the facts by asking. Check the facts by observing. Check the facts by stepping back from a situation before you take action. Try to see a whole or a bigger picture, or even put yourself in the vantage point of somebody else."

There is a popular saying, coined by Robert Heinlein: "Pay it forward." Well, Dialectical Behavioral Therapy has this concept in relation to emotional modulation: "Cope ahead." In other words, rehearse what it is that you can do when you find yourself in difficult situations.

I see a lot of clients for fear of public speaking. I live in an area where of the country where there are a lot of oil companies that employ people who are both geologists and attorneys. What's interesting about them though is that for the most part they have been writing scientific journals about oil and gas exploration, or they have been writing policy related to environmental concerns from a legal perspective. Even though they're licensed as attorneys, few of them ever are in a courtroom. A few of them ever have to testify in public. Most of those whom

I've worked with are quietly writing documents and other things for the oil companies where they work.

Years ago I actually had one of them call me up and say, "I've been a lawyer for twenty-five years, and I've never been in court. Now, on behalf of the oil company, I'm actually having to appear in court, and I'm afraid to speak in public." This is a guy with a Ph.D., who's an expert in his subject. So I worked with him to overcome his fear of public speaking. Since then I've built a large number of referrals particularly in the oil and gas industry: executives and lawyers who for one reason or another have a fear of making presentations, testifying before Congress, trying a case in court, defending their positions in a public forum, or whatever it is that they need to do.

With almost all of these clients, I guide them through a process of coping ahead, of visualizing themselves succeeding. This really is the same technique that coaches use in athletics to help a person experience winning the game before the game takes place. If I can help my client to experience success in a future problem, I can help my client to modulate distressing emotions. This is called by many different names, including Future Pacing (in hypnosis and NLP), Cope Ahead (in Dialectical Behavioral Therapy), and Mental Rehearsal techniques (in coaching and personal fitness training).

Finally, problem-solving skills contribute to emotional modulation in DBT. After all, if a distressing problem is resolved, it is no longer distressing.

To sum up, here is a list of the Emotional Modulation skills in DBT:

+ Understanding Emotions
+ Learning How Emotions Function
+ Experiencing Emotions without Overwhelm
+ Reducing Emotional Vulnerability (HALT)
+ Building Positive Experiences
+ Checking the Facts

- Coping Ahead
- Problem-Solving

Distress Tolerance Skills

In Dialectical Behavioral Therapy, there are really two types of distress tolerance strategies. First of two types is acceptance. We've talked about that a great deal in Acceptance and Commitment Therapy, and really the ideas in DBT are fairly similar.

In DBT, the idea of acceptance is developing the ability to accept the situation and oneself without judgment. This strategy assumes that emotional suffering is a part of life and that avoiding this reality leads to increased emotional suffering. Acceptance is learned through breathing and awareness exercises.

Belly Breathing is one of my favorite techniques to teach clients who are experiencing distress. While it's easier to teach this in person, we can still explain it here in this book:

Just take in a deep breath right now. Take in a deep breath; breathe in, and now exhale. Now, I can't see you, but chances are good that you just did what I did the first time I tried this exercise. Chances are pretty good you puffed up the chest a bit. You breathed through your nose. When you exhaled, it's likely that your body collapsed a little bit.

What's interesting about that is that we tend to almost always breathe like that when we take a deep breath. When we're told to take a deep breath, we puff up that chest. We push that neck back a little bit. We breathe in.

But when we breathe with the chest, we actually only get air to the top part of the lungs. We don't get oxygen in the lowest part of the lungs. We also tense up those muscles in the neck and shoulders. That's where the collapsing comes from.

So, I'm going to teach you a technique that you can teach your clients. If you're not already, sit comfortably in a chair with your back and spine aligned and erect. You don't have to puff out the chest at all. Just let your arms rest to the side for a moment and pay attention to the belly. In fact, you can even rest your hands and put them across the belly if you'd like.

Now, imagine that inside of your belly is a balloon like a kid might have, a healing-filled balloon, but imagine there's no air in it. It's just an empty balloon inside of your belly. Now, the way I want you to breathe right now is to breathe by imagining that you're filling that balloon up with air. Go ahead breathe in. Imagine that you're filling that balloon up with air. And notice how the chest stays the same and the belly fills up. Now exhale by just letting the air out of the balloon. Again, breathe in by filling up that balloon. Fill it up with air all the way in and then exhale by letting the air out of the balloon.

Do you notice the huge difference? Chances are you feel wonderful from that burst of oxygen that you've just experienced going all the way to the lowest part of the lungs without increasing any tension or stress in the shoulders or neck.

That's a simple strategy, but it's a strategy that I teach to every one of my clients who is quitting smoking.

Almost all of my clients say, "Yeah, in the morning I need a cigarette or two to get going." Nicotine is a vasoconstrictor; for them it's a stimulant that provides energy. So they are probably dependent on nicotine for their energy just like other clients are dependent on carbohydrates. Since they don't have a cigarette anymore and need that energy, I teach my client to start the day with a couple of breaths like this because when oxygen goes to the deepest part of the lungs and then goes through the heart and through every cell of the body, they get a blast of energy as powerful and more powerful than that produced by the drug nicotine.

Besides acceptance skills, crisis skills are important for our clients. These strategies focus on finding new ways to survive and manage the moment without resorting to problem behavior. There are really four sets of crisis survival skills that clients learn in DBT:

- Distracting
- Self-Soothing
- Improving the Moment
- Thinking about the Pros and Cons

Distraction Methods

Distraction Methods are really based on the idea of paying attention to something other than what is causing the distress. This is of course a factor in Mindfulness: When you are paying attention to the breath, you are no longer paying attention to your distress.

Acceptance strategies for distress tolerance are usually addressed within the Mindfulness techniques that we've discussed throughout this course, but Dialectical Behavioral Therapy uses an acronym to tie together the crisis management strategies that incorporate acceptance principles. The acronym used is ACCEPTS:

- Activities
- Contributing
- Comparisons
- Emotions
- Pushing Away
- Thoughts
- Sensations

Activities: Clients are encouraged to distract for themselves with pleasant activities, hobbies, going for a walk, cooking, gardening, watching a movie, playing a sport. Have your clients write down a list of pleasant activities that they like to do. Borrowing from ACT therapy, ask them to commit to the valued path to engaging in one of those activities when they find themselves in crisis.

Contributing: Contribute to others to the world around you. By focusing on helping someone else, your clients may often find relief from their own problems. This is one of the premises again of Alcoholic Anonymous. They say if you're new to a meeting, what should you do if you're a newcomer? Well, you should stack chairs and empty ashtrays. Similarly, my musician friend James Hazlerig found that he often became sad and lonely at the end of music festivals, so he began focusing on helping other people pack up their camps as the festival was drawing to a close.

Comparisons: Your clients may benefit from comparing how they're doing now as opposed to one or five years ago. (Often, simply being aware of the problem is a big improvement.) They may also compare their own problems to those of people facing monumental disasters. After all, being stuck in traffic seems pretty minor when compared to being stuck in a hurricane.

Emotions: Clients may scale their emotions to gain perspective, or may decide to pursue an emotion opposite of what they are feeling at the time. For instance, clients may listen to happy music when they are feeling down. By the way, especially if you work with adolescents, find out what they're listening to and change their play list. That doesn't mean they have to stop listening to that sad, depressing music they love. They can still do that when they're doing fine, not when they're experiencing crisis.

Push Away: Rather than be constantly confronted by a distressing situation, clients can learn to "push it away" and come back to it later. For instance, when it's time to go to work, a client may visualize putting problems into a box until after work. Or clients may push away a situation until they are emotionally and physically prepared to deal with it. Mindfulness exercises are very helpful for developing this skill. Of course, it may not be healthy to avoid problems forever, but this strategy helps clients to function without being consumed by their distress.

Thoughts: Clients learn to distract themselves with thoughts other than negative and distressing ones. They may count to ten or focus on the stars in the sky, anything that keeps them focused away from their negative thoughts. This is a useful strategy in crisis situations when the client needs to access a strategy to manage the distress quickly. Other thought distractions include reading, watching movies, writing a journal, and doing crossword puzzles.

Sensations: There's a saying, "If you want to feel something different, *feel* some *thing* that is different." Clients are taught to distract themselves form emotional turmoil through other bodily sensations such as holding ice or flicking a rubber band on their wrist. These sensations produce shock and can distract from emotional pain. These sensation exercises are more helpful ways of managing distress that cutting and other self-injurious behavior. Other sensations could be listening to loud music or taking a cold shower. Following a sensory distraction, the client may then engage in an activity distraction.

Self-Soothing

I teach my clients body scan meditations, similar to those call Shavasana in yoga classes. Progressive muscle relaxation is another basic strategy that can be used to teach people self-soothing techniques.

Improving the Moment

Sometimes a client comes into my office and they're experiencing crisis, so I ask them this very simple question: "I recognize that you're in crisis, and that's painful and difficult. Is there anything you can do right now to improve the moment?"

That's something they've never thought about before. Sometimes that question catches them off guard, but they will often come up with an answer, if I'm quiet and patient. (By the way, one of the hardest things for therapists to do is to be quiet and wait for the answer. I've had to learn that as a strategy in my therapeutic interviewing, but it's a strategy that's been really helpful to me.)

Sometimes clients come up with some of the ACCEPTS strategies, and sometimes I suggest some of those strategies. But another strategy for tolerating negative emotion is to improve the moment by staying mindful.

IMPROVE is another acronym in Dialectical Behavioral Therapy and it is used to remember these skills:

- Improve mental imagery.

- Meaning: Find that which valuable to you in the moment.

- Prayer: a useful way within the context of our client's faith to be in the moment.

- Relaxation: Relaxation techniques are extremely useful and helpful to our clients. Relaxation is the basis of developing the skill of taking physical control over the emotional. Progressive muscle relaxation and autogenic training in particular are two excellent skills for doing that. I'm surprised at how many therapists do not teach clients progressive muscle relaxation, autogenic training, and self-hypnosis. The efficacy of these approaches is extremely well-documented. Many of you may remember Herbert Benson, a Harvard psychiatrist who in the 1970s wrote a book called The Relaxation Response, documenting the many benefits of deep relaxation.

- One thing at a time.

- Vacation: maybe not a real vacation, but a mental vacation. Step outside for a moment. My back patio at my house is my place for a vacation. Sometimes I decide I'm just going to take the laptop out there and work, or maybe work out there on my porch. A couple of weeks ago, my 6-year-old step-daughter and I decided to go outside, make lemonade with fruit in it, simply sit out there, and have a little mini vacation—and that's exactly what we did. It was fun, and the lemonade was good, too.

◆ Encourage positive self-talk: Do you remember the little engine that could? I think I can, I think I can, I think I can. I want my clients to use positive affirmations as described in earlier chapter, and that approach fits well within the ideals of Dialectical Behavioral Therapy.

A Relaxation Exercise

This is a script you can use with your clients, and of course, I urge you to read through it several times so that you can then guide yourself through the process.

Find a comfortable spot where you can relax. You can lie on the floor. You can sit in your chair. This session is not designed to help you sleep. It's designed to help you use the principles of relaxation to help you end any feeling of anxiety, anger or depression.

You can come back to this often and you can learn how to practice these principles by going through this exercise in a future date as well. Many people who use the principles of relaxation find that by taking a moment in the middle of life's turmoil to re-center themselves. It's a great way to relax. And you might even wonder can something so simple really help you and the answer is yes. There is hope for ending depression, and it comes from the desire to be happy that's already within you. You can find calm from anxiety, and you can even give up stress, anger or frustration by really practicing these simple ideas.

And so as you relax, close your eyes and picture in your mind a relaxing waterfall or a moving pool of water. Then simply scan your whole body, and anywhere that you're holding the tension of the day, let those muscles become lose and relaxed. Often we don't even notice where we're carrying the tension until we notice it and make a conscious choice to relax those muscles.

As you mindfully focus your attention on the creative part of your mind to imagine a waterfall, let the water flow and let any immediate stress or anxiety flow with the water from you and into a foreign distant place. And imagine for a moment the feeling not only of the water moving, but also your stress moving with it.

Many will find that easy to do, but others won't find this an easy thing to do. Perhaps that's because the stress has been a part of life for so long and letting it flow is something that takes time for some people, and that's perfectly okay as well. But notice how by just using this visualization your mind has become more relaxed. Your bodies become more relaxed. Your eyes have become tired and heavy. In fact, you'll find that even though you know you could open your eyes if you wanted to, it just feels so good to keep them closed. You'll just let them stay closed as you continue to use your imagination to drift into the scenery of a waterfall or pool of water or calm sea.

Now, notice your breath, smooth and rhythmic. It isn't even something you've tried to do, but something that has come naturally to you by taking a moment for yourself and letting go of any obvious tension. And as you rest your hand on your lap, notice the sensation of relaxation and calm you've already achieved, and say to yourself the word "warm." Focus on your hands letting your hands feel a sense of warmth. Create that warmth and say to yourself, "My hands are warm."

And as you do, notice the sensation of warmth in those hands that you've created and now say the word "heavy" and notice how heavy your resting hands are, and say to yourself, "My hands are warm and heavy; my hands are warm and heavy."

Now, focus on your feet, saying to yourself, "My feet are warm and heavy; my feet are warm and heavy." It's amazing how as you say this, you can begin to feel a sense of warmth from within and a sense of heaviness.

This is really your first learning in this relaxation exercise that you can control the way you feel. It's true of physically and mentally, no matter how difficult life's situations are. And that sense of warmth and heaviness brings a sense of relief from the weariness of life, letting you recharge the mind and body in the time that we have in this session.

Now, again, notice your breathing, smooth and rhythmic, relaxing even deeper with each breath, and although it might feel magical to relax as deeply especially since you haven't felt this calm in a long time, this is a totally natural state.

It's one that you have created rather than one I've created, and because of that you can re-experience it at any time, and I will teach you how to do that in a moment. But right now focus on your desire to be free from panic or anxiety by feeling a state of calm. Notice how the sense of heaviness is like a calming anchor allowing you to feel physically calm even if your mind might wander or race. Perhaps you desire hope rather than depression, and there's a simple exercise of creating warmth and heaviness.

You can see how you have the ability within you to not only create those sensations but also a state of hopefulness. I don't know why, but water always makes me feel hopeful, maybe because it reminds me that nothing stays the same or perhaps because of its power and energy to revitalize my spirit.

Do you desire freedom, success? Take a moment in creating affirmation that reflects your heartfelt desire. You could say something like this: "I'm free from stress" or "I can create hope from inside."

You could even focus on a single word like happiness or calm or forgiveness. It's amazing how creative our minds are. You can actually hear the water in the background as you take a moment to focus on that word: happiness, calm, forgiveness, or that affirmation that you've created. And repeat it in your mind, sing the letters as you do, spelling it out, hearing yourself speak those words and noticing the feeling that these positive words or affirmations bring.

It really is amazing how simply saying something manifests it as reality, but remember everything that is now was a thought first. To be hopeful or calm and forgiving, you must begin with a thought, and you can truly experience this more and more each and every day.

A great way to reinforce this thought, which like rain fills the stream and leads to the river and results in the formation of a powerful waterfall, is to write that affirmation on a card or a sticky note and tape it to the bathroom mirror or your dashboard or to the monitor of your computer.

Let it be a constant reminder despite other factors in life of your ability to turn your thought, your mental energy into emotions and success, and relax even deeper noticing how remarkably easy it's been to set aside a few moments to re-energize, and even though life requires action you now have a starting point for re-energizing during difficult times and a starting point for creating thoughts which turn into results. In fact, you can even congratulate yourself for taking the time to invest in your success by starting this process of hope, calm, and happiness.

Continue to relax. Enjoy just another moment or so of this state that you've created, never asleep, but deeply relaxed, enjoying your ability to experience serenity and creativity, and to let any temporary stress pass without escalating.

As we end—near the end of our time—you can open your eyes at any point, or you can keep them closed for another moment or two where you feel the floor below your feet and the air in the room around you. Maybe even take another moment to continue to see or hear that waterfall that you've created, but since it's time to conclude, begin to feel a sense of energy to let the muscles in your body and your spirit feel energized. If your eyes aren't open yet, it's now time to open the eyes feeling refreshed, energetic and ready to be positive in every situation in every way each and every day.

One, two, three, eyes open, feeling fantastic.

That is a simple process of relaxation training that incorporates a number of different strategies including direct suggestion, progressive muscle relaxation, autogenic training, and a number of other approaches.

There are a number of other strategies which I utilize with my clients all of which are predicated on helping them to feel awesome, not only at physical level but an emotional level as well. My hope is that you will actually take the script and use it with your clients so that they can derive the benefits of these evidenced-based protocols as well.

CHAPTER ELEVEN

Metaphor and Story

One of the things that's interesting about Contextual Psychology is that the value of metaphor and storytelling is recognized across traditions within this arena, although in truth, this isn't something restricted to Contextual Psychology.

I remember when I was first learning how to be substance abuse counselor, my first job in this profession. I was at an old-fashioned treatment center complete with chain smoking chemical dependency counselors and orange vinyl furniture. Every new client was put on the hot seat. It was before managed care, and it was assumed that everybody magically got well on the 28th day, nobody on day 29 and nobody on day 27.

At that center, there was an old-school counselor named Sam who had quite a reputation for grilling clients in the hot seat. Graduates of the program often got bumper stickers that read, "I survived Sam."

Now, Sam was somebody who recognized the value of the stories in the big book of Alcoholics Anonymous. In fact, most of the big book is taken up with stories that really teach some important lessons. Sam would regularly tell clients those stories from the big book of AA because he knew the value of the story in therapeutic counseling. It was perhaps because of my experience in substance abuse counseling that I recognized right off the bat that stories are powerful agents of change.

I love to tell stories to my clients, and I do so on a regular basis. One of the reasons why is I'm not very good at confrontation. Some of those old time substance abuse counselors would put a client on the hot seat and get up on their face like a bad Fritz Perls video. They would even swear at them and they would make them "get in the now" as we used to say back then. It's just not my personality to get up into somebody's face and just tell them like it is. I'm more diplomatic than that.

And I remember thinking to myself, "I'm not going to be a good substance abuse counselor like Sam because I can't get up in their face. I can't make them get in the now. I can't be six inches away from them like a drill sergeant and confront, confront, confront."

However, I recognized that, even though my personality was certainly different from Sam's personality, I could be an effective substance abuse counselor, that I could be effective at confrontation, if I learn to use stories as a way of confronting clients.

Here is one of my favorite stories. It's paraphrased from Leo Tolstoy:

A bishop and several pilgrims traveling on a fishing boat from Archangel to the Solovetsk Monastery. During the voyage, the bishop engages the fisherman in conversation after overhearing them discuss a remote island nearby and he asked him to change course because this island nearby has three old hermits living a Spartan existence.

These old hermits are seeking salvation for their souls, and several other fishermen claimed to have seen them once. The bishop then informs the captain that he wishes to visit the island, and the captain attempts to dissuade him by saying the old men are not worth your pains. I have heard it said that they are foolish old fellows who understand nothing and never even speak a word, but the bishop insists, so the captain steers the ship towards the island, and subsequently sets off in a row boat to visit where he is met ashore by the three hermits.

The bishop informs the hermits that he has heard of them and of their seeking salvation. He inquires how they are seeking salvation in serving God, but the hermits say they don't know how. They only pray simply: "Three are ye, three are we, have mercy upon us." Subsequently, the bishop acknowledges that they have a little knowledge, but are ignorant of the true meaning of doctrine and how to properly pray. He tells them that he will teach them not a way of his own, but a way in which God and the Holy Scriptures has commanded all men to pray to Him, and then he proceeds to explain the doctrines of the incarnation and the Trinity.

He attempts to teach them the Lord's Prayer, the Our Father, but the simple hermits blunder and cannot remember the words, which compels the bishop to repeat the lesson late into the night. After he becomes satisfied that they had in fact memorized the prayer, the bishop departs from the island leaving the hermits with a firm instruction to pray as he had taught them. The bishop then returned by the row boat to the fisherman's vessel anchored offshore to continue the voyage.

While on board, the captain notices that their vessel is being followed—at first thinking a boat was behind them but soon realizing that the three hermits had been running across the surface of the water as though it were dry land. The hermits catch up to the vessel. As the captain stops the boat, the hermits inform the bishop, "We have forgotten your teachings, O Servant of God. As long as we kept repeating it, we remembered, but when we stopped saying it for a time, a word dropped out and now it has only gone to pieces. We can remember nothing of it. Teach us again."

The bishop was humbled and replied to the hermits, "Your own prayer will reach the Lord, men of God. It is not for me to teach you; instead, I ask you to pray for us sinners." The hermits then turned around and walked back to their island.

This story by Leo Tolstoy has long been one of my favorites not just because it is a classic in literature, but because the story teaches so many valuable lessons. That's really why a story should be told or used in therapy: because it's a method of teaching and a method of confrontation.

Story Elements

Let's talk about the elements of a story. Pretend this is a literature class for a minute. What is a story? It's an imaginary or a real account of people and events. There's a protagonist, a main character, a hero with whom the audience usually identifies, and this is really important because when we tell stories to our clients, they see themselves in those stories. One of my favorite stories from the big book of Alcoholics Anonymous is a story of Jim the car salesman. Jim because of his own alcoholism ends up working as a salesman for a car dealership that he used to own. Many of my clients, even if they've never owned or worked in a car dealership, can relate to the main character in that story because they too have sometimes suffered humiliating consequences because of their drinking.

Every story has three different types of scenarios in it: an antagonist or a challenge or a conflict. And every story has a resolution. Stories are great because every story is ultimately about change. If there is no change in the story then there really is no story. Stories have themes and morals. The message of the story may be explicit or it may be implicit, but it's still there.

Metaphors make excellent stories. One type of metaphor is a comparison or analogy stated as equivalence, for example Road Hog, Couch Potato, Rug Rat. And parables are something we're all familiar with both from Christian Scripture as well as Eastern Scripture: a story or short narrative designed to reveal allegorically some principle, moral lesson, psychological reality, or general truth.

Therapy of course is an activity or interaction intended to bring about rehabilitation or social adjustment, and stories, being about change, are excellent tools for accomplishing those goals.

Stories are natural hypnotic inductions, natural trance inducers. Now, when I mention trance, people often think of Hollywood images of Igor mindlessly following Dracula, holding his arms out blindly like a zombie walking through the night. That's not what trance is. Trance is a state in which internal perception has become more important than external perception or in which a limited aspect of external perception has become so important as to preclude the rest of the external perception.

And trance happens naturally through the process of storytelling. My mother is a Montessori preschool teacher. When I observe her classroom, kids will be all over the room doing their own thing in their own different corners, but sometimes she brings them to the line. The line in a Montessori classroom is where kids learn things, and sometimes on the line she will tell them a story. Even though they've all been doing different things, they become a single focused group during story time, their eyes all watching my mother as though they were mesmerized.

It's amazing how kids who are so busy doing other things can set all of that aside to enter into a trance state where internal perception—the story they are imagining as it is told—has become much more important than external perception. So, trance is not something unnatural. It's not something mysterious. It's something we all experience on a regular basis, and storytelling is a great way to induce trance in therapy.

Why would we tell a story or a metaphor in therapy, for example, the story of the old hermits of Leo Tolstoy? Well, stories shift the filters of our perception. They have unconscious meaning. They bypass conscious resistance. If I tell somebody what they should do, there is a power struggle, but if I tell them or share with them a story, it bypasses conscious resistance to the message. It is a great way to conceal a confrontation.

As we've been talking about in ACT therapy, a story can break old relational frames and create new relational frames. I love stories that are based on metaphor, and ACT therapy in particular recognizes the value of metaphor. A metaphor is like a seed. It plants an idea in our unconscious mind. It germinates in the subconscious mind and comes to action in our conscious mind.

Here are some examples of short metaphors:

- Life is a journey.
- Purposes are destinations.
- Means are routes.
- Difficulties are obstacles.
- Counselors are guides.
- Achievements are landmarks.
- Choices are crossroads.
- A lifetime is a day.
- Death is sleep.
- A lifetime is a year.
- Death is winter.
- Life is a struggle.
- Dying is losing a contest against an adversary.
- Life is a precious possession.
- Death is a loss.
- Time is a thief.

I mean there are just so many different metaphors and each one of those metaphors can be very powerful and activate subconscious learning. It bypass critical or conscious resistance and our clients will attach unconscious meaning through these things.

Passengers on a Bus

As I mentioned, metaphorical stories are an important part of ACT therapy. A group of ACT students have actually made short animated videos of some of these stories; they can be found on the Internet and are easily shared with your clients: http://youtu.be/Z29ptSuoWRc

One of the earlier ACT metaphors, and one of my personal favorites, is called "Passengers on a Bus." This could be told to a client in either of two ways:

- ◆ Explicitly: telling them what the metaphor is all about
- ◆ Implicitly: letting them attach their own meaning to it.

Let me share with you an introduction to both ways. Explicitly, I might say this:

I'm going to share with you a story about being a bus driver. And when I talk about picking up passengers at different bus stops, this is really like our mind picking up thoughts.

Then I can go into the story with the meaning already explicitly stated.

Alternatively, I could let the meaning be implicit:

Let me share with you a story. Imagine that you are a bus driver. [I like to make the client the main character.] *Imagine that you're driving your bus, and as you drive your bus, you stop at various bus stops, to pick up passengers.*

Notice I've begun the story just by setting up the story, but I haven't told them what it's all about. In implicit storytelling, I trust that my client, through his own or her own reservoir of knowledge, will attach the meaning and significance to the story that's most important to them. That really is very similar to the tradition of Milton Erickson, who is one of the master storytellers in the field of psychotherapy and family therapy in particular.

Imagine that the chair you're sitting in is the bus driver's chair. In fact you can even hold your hands up if you want to as if they're on the steering wheel and imagine that you're driving the big bus.

As you go to the first bus stop, you're going to be picking up some passengers. Of course, you didn't choose the passengers that ride the bus. They choose to ride the bus, and they might not necessarily be the passengers that you want on the bus. Maybe one of those passengers is called self-doubt. Maybe one of those passengers is called criticism. Maybe one of those passengers is called despair. But as you drive the bus, you're going to be stopping at another bus stop and picking up even more passengers.

And again, some of these passengers may be passengers you enjoy, maybe a passenger of joy or a passenger of calm. But maybe you'll also pick up a couple of other passengers you don't really want on the bus. And as you continue on to the third stop and the fourth stop, picking up more and more passengers, the bus begins to fill. And as you drive the same old route picking up passengers, you noticed there are other places you could go. You could go the scenic route, or you could go the interesting way, or of course, you could just continue on the similar route.

If you decide to make a turn and deviate from the same old route, maybe some of those passengers will speak up. They'll start talking to you. Maybe self-doubt will say, "Are you sure you want to do this?" Maybe criticism will say, "That's not the best way to go." Maybe anger will say, "Hey, get back on the same old route. We don't want to be over here."

Some of those passengers might actually have something nice to say: "I'm really glad that you decided to take the scenic route today. I'm really enjoying the interesting way today."

As the bus driver, it's impossible to listen to the direction from all 60 people. So, you of course get to pick and choose who it is that you'd like to listen to. Or, maybe as the bus driver, you've decided that they're simply passengers on a bus and you don't have to listen to any of those passengers. You can simply drive the bus wherever you would like.

Now, there's an example of a metaphorical story from ACT therapy.

The Tomato Story

Milton Erickson, a psychiatrist and family therapist who revolutionized hypnosis and has had a huge influence on modern therapy, told a lot of stories, including personal stories, in his therapeutic work. One of my favorite stories is a story that Milton Erickson told, but I tell it from my own perspective because I think that Milton Erickson and I must have had the same grandmother. In fact, I'm going to share this story with you. Perhaps you even had the same grandmother I had or the same grandmother Milton Erickson had. I often ask my clients to close their eyes and just relax as I tell this story:

When I was a kid, my grandmother had a garden, and I'd love to go out in the garden and help my grandmother work. It was one of my favorite things to do as a kid.

My grandmother always had flowers in her garden. She had vegetables in her garden as well, and she even had fruit trees. Growing up in Chicago, I always look forward to the melting of the springtime snow when my grandmother could plant the seeds for the new year. She would take two stakes and a string and stretch it out and plant her seeds. Now, I remember I would just lie there on the earth and watch each and every day as my grandmother cared for her plants and cared for her seeds, waiting for those seeds to begin to germinate.

It was always exciting when they would start to come out through the earth, and they would begin to grow, and the vines would grow a little bit longer and a little bit larger, and the vines would begin to have little yellow flowers on them. I would watch those flowers until eventually they would melt away, and a small green bulb would emerge. That green bulb each and every day seemed to get a little bit bigger and bigger and bigger. My grandmother would prune the garden, and she would water the garden, and she would clean anything out of the garden that had made its way in there that shouldn't have been in there. She would fence the garden to keep the rabbits away.

She cared for her garden each and every day, and each and every day those green bulbs became larger and larger until eventually they began to change color: a little bit orange and eventually red. And by the end of the summer, my grandmother's garden was yielding the most beautiful and delicious tomatoes I've ever eaten in my life. In fact, when I was a kid, one of my favorite things was before dinner when my grandmother would go out in the garden and pick some of those tomatoes, and she would cut them up and put them on a plate with a little tiny bit of oil and a little tiny bit of balsamic vinegar. She'd add a little bit of sea salt and a whole lot of pepper. We would have those fresh tomatoes before a meal as a salad or an appetizer.

It occurred to me one day long after I was an adult those tomatoes never had to ask my grandmother, "How much rain from the sky should I drink? How much nutrition from the soil should I absorb?" Each and every spring my grandmother would plant the seed. She would tend her garden, and the tomatoes never spoke to her. They never asked her anything. The tomatoes simply did exactly what a tomato is supposed to do, yet by the end of the summer, they were absolutely the perfect size, the perfect texture, the perfect color, and completely delicious.

What's the point of that story? In this story, I was mostly implicit rather than explicit. Whom could that story help? Remember, I work with obese clients. When I tell the story to my obese clients, I usually tell it with long, lean cucumbers rather than tomatoes. But I tell the story with tomatoes as well on a regular basis. What does this tell an obese client? *Listen to your body. Our bodies will tell us what we need.*

What other type of client might benefit from this type of story? Think about it for a moment. There are a lot of different messages that our clients could attach to such a story, and stories like these are excellent things to share with our clients to teach them both implicit and explicit truths.

Milton Erickson shared wisdom in his stories. He told stories and metaphors to confront and he used stories as a way of making suggestion.

In addition to ACT therapy and Ericksonian hypnosis, where are stories used? Well, of course, the big book of Alcoholic Anonymous. It actually starts out with a story and a poem about the war. Likewise, Jung in psychotherapy certainly was predicated in large part on the mysticism of stories. Pastoral counselors in their sermons are often using stories, the parables of Christ.

Why are Stories Therapeutic?

In a story, there is an identity. You become the story. In a story, I can see a bigger picture and how my experiences might be related to others. For example, the story of my grandmother I told that from my personal perspective. What's the application here that there's somebody who cares about you that will take care of you. Maybe the lesson is there's somebody who can teach you. The therapeutic value of a story is to help clients choose a valued path (as they say in ACT therapy), to adapt morals that are consistent with the values that are most important to them, and in many cases to teach self-control. (Many of Aesop's fables are about self-control.)

Stories provide a vicarious experience. Just last night as I was setting up a new laptop that has Windows 8 on it, I noticed that Microsoft provided me with a tile called "Travel." Although I have had the opportunity in years gone by to travel all throughout the world, it has been a number of years since I've done much traveling, so I noticed that tile and pressed the button. I kept scrolling through and looking at all of the places. For about an hour I just enjoyed this webpage provided by Microsoft and had a vicarious experience that didn't require any airfare, air travel, international hotels, passport controls, or bed bugs. Stories do the same thing. They can provide a vicarious experience, which can be particularly useful both to couples and individuals.

If you choose to implement storytelling into your repertoire of techniques and strategies in therapy, there are a couple of cautions, and one of those is only tell stories you're familiar with. If you want to tell a new story, simply practice it. One of the interesting things about therapy is that we don't think we have to practice it. A musician practices his instrument. A singer practices his singing. A hockey player practices his hockey sticking. A basketball player practices his free throws, but in therapy school we're never told to practice our therapeutic technique. Yet we need to practice in order to be successful. So become familiar with stories. Make sure you know the story and practice those stories.

If you are telling stories to clients and letting the meaning be implicit rather than explicit, make sure the client whom you're working with has abstract reasoning skills.

Of course, it's important not to break rapport by telling a story. Storytelling should only be told to increase rapport, to engage your client, to draw them in. Tell a story clients can relate to if you choose to tell a story. For example, I never tell my clients religious stories unless they have already told me that those stories are important to them.

Not too long ago, I had a smoking cessation client. He needed to quit smoking because he was not healing well after an amputation, and of course smoking reduces circulation. It causes complications in healing from those types of surgeries, so when he came to see me, he was in really bad shape. I didn't know where to start with this guy, but on my intake form, he had written that he liked Bible stories.

It's the only thing he had written. For those who don't know, my bachelor's degree is in ministry, and much of my education has been religious education, so I'm very familiar with Bible stories. Because he brought that to my office, I shared with him the story of creation in Genesis, in which God creates by speaking. And then I moved into affirmations so that he could create the state of being a non-smoker.

I tried to tell stories my clients can relate to. Maybe if I have a client who's got a Masters Degree in English Literature or a Masters Degree in Russian Literature, I'm going to share Leo Tolstoy's story, or if I'm talking to somebody who's about my age and they tell me their favorite thing was watching Saturday morning cartoons, I might tell them one of Aesop's fables.

If you are telling a story, don't get the details wrong. It's important not only to remember your story, but also to make sure that you're a good storyteller who can cover all of the details. By the way, there is nothing wrong with reading a story to your clients. Absolutely nothing wrong with it. In fact, I keep a number of story books on my shelf, and even though they're stories I'm familiar with and have read many times, I'll sometimes say to the client, "Let me read for you a short story." And I'll actually pick a book and read a story, even one out of Alcoholics Anonymous.

Where do you find stories? They're everywhere: mythology, folks tales, dreams, movies, fables, parables, TV shows, history, life, Milton Erickson, Chicken Soup for the Soul, the big book of Alcoholics Anonymous. A story doesn't have to be true. It doesn't have to be perfect. It can be a joke. It can be humorous. It can be funny. However you choose to tell a story, recognize that that story though can be quite valuable to your client.

Six Steps to Good Storytelling

According to my friend James Hazlerig who is a master storyteller in Austin, Texas, there are six steps in good storytelling. He has a Masters Degree in English Literature, and he earns his living telling stories at Renaissance Fairs. He's also sees clients in the Austin area. The Six Steps are as follows:

1. Learn to enjoy and collect stories. Every time you go to the bookstore and you think you're going to buy an important self help or therapy book, also buy a story book or a poetry book.

2. Know the structure of a story. Every story has conflict and resolution. Most traditional stories use three events to establish a pattern.

3. Adapt and improvise on a story if needed to meet your client where they are, for their particular needs.

4. Make eye contact when you tell a story.

5. Use the voice as an instrument. Find your rhythm and cadence, using tonality that punctuates, and always remember that silence can be as valuable as a spoken word in telling a story.

6. Show a story with your words. Don't tell it. Use sensorial descriptions: sights, sounds, scents, touches, flavors.

Metaphor, story, parable—all can be effective components of therapy, and all are employed in Contextual Psychology. It's my hope that this chapter has inspired you to collect a repertoire of therapeutic stories to use with your clients.

CHAPTER TWELVE

Case Aplication and Structure
of Therapy

Let me share with you two cases in which I used the methods of Contextual Therapy over the last couple of months. Of course, I've changed the clients' names and other demographic information to preserve confidentiality.

Suzie's Case

Suzie is a 35-year-old female who came to me with presurgical anxiety. In fact, when she called she basically said, "Look, I need to see you as soon as possibly can because I'm going to be having surgery two weeks from now. She got my number from another client whom I had helped actually sometime before.

But she didn't hear about me until just a couple of weeks prior to surgery. What I had time to do during those two weeks was three sessions, and this is very important. Most therapists traditionally schedule sessions one week apart. Why? Because that's what therapists do. For how long? For fifty minutes. Why? Because that's how long a therapy session is. I almost always spend at least ninety minutes with my clients in a first session, and I'm not an avid clock watcher. Sometimes I will spend thirty minutes with a client in a follow up session, and other times I'll spend another hour and a half when they're in a follow up session. That really depends on the client and what the presenting issues are. Rather than trading my time for dollars, I sell solutions.

So I scheduled three sessions with her in about a ten-day period. I recognized that I only had three sessions to work with her. Because anxiety was the number one issue, I spent my time teaching her first-session, second-session, and third-session core skills of Mindfulness Meditation and exercises of Acceptance.

Earlier on in this course, I took her through an exercise where a person pictures themselves sitting in a conference room table. No chairs are around. It's just them and a box in the middle of the table. They could take those things out of the box, put them on the table, and simply observe that they're present. I used that imagery with this particular client who was having a very hard time accepting the physical condition that she was in and the experiences she was about to have being thirty-five years of age.

Notice I did not ask her in what other areas of life she was experiencing anxiety. Because three sessions is not a long period of time, I kept our work tightly restricted to the immediate problem: Twelve days from our first meeting she was going to be having a surgical procedure, and she was scared out of her mind. And that fear might have caused her to make decisions that are unhealthy, maybe even in the extreme canceling the necessary surgery, which was something she was contemplating.

And so by using a combination of Acceptance exercises plus basic Mindfulness training in that first session, she was able to experience at least the hope that she would be able to manage her anxiety better between then and the surgery. In session two, she was still having a lot of physical symptoms associated with anxiety, even though she reported that the Mindfulness exercise was in fact tremendously helpful to her. In fact she was one of those clients who found it so helpful that instead of practicing with intention twice a day as I often ask my clients to do, she said she practice three or four times a day.

Now, it was probably four or five days after the first session that I had the second session with her. She certainly was improved, but anxiety was still so bad that it was making her physically sick. So I taught her Progressive Muscle Relaxation, which was pioneered by Edmund Jacobson in 1928. I think almost every therapist learns about Jacobson's Progressive Muscle Relaxation at some point in their career. The research validating the effectiveness of Progressive Muscle Relaxation is incredible.

I did not take forty-five minutes to go through a process of Progressive Muscle Relaxation with her. I didn't go through all sixteen of Jacobson's different muscle groups. What I did was what I do with most of my clients when I teach them Progressive Muscle Relaxation. I teach it in a very non-threatening and simple way, and I usually take about ten minutes to go through the basic process. I use a process of passive relaxation rather than having them tense and relax, tense and relax each muscle, the way the Jacobson did. I do however start out with a tense-and-relax cycle for the fists before starting at the top of the body and going through the various muscles of the body: the shoulders, the arms, the belly, the back, the seat, the thighs, the shins, the calves, and even the little muscles of the feet.

I find that Autogenic Training blends well with Progressive Muscle Relaxation, so with this particular client, I taught those strategies as well in the second session.

In my third session with her, I reviewed her practice. I asked her some solution-focused questions from Solution-Focused Brief Therapy, and then I had her future-pace her wellness. What is future-pacing wellness all about? It's all about really what I talked about in the Positive Psychology. Creating a thought that produces an experience. I gave her homework assignments for after her surgery: some MP3 Autogenic Training and Progressive Muscle Relaxation sessions that I have prerecorded. Her instructions were to continue practicing these things after the surgery was over because the research shows us that these strategies of Mindfulness, Progressive Muscle Relaxation, and Autogenic Training when talked to clients before surgery and utilized after surgery do three things: reduce complications, decrease dependency on drugs, and increase the speed of recovery.

Bob's Case

Bob was a 56-year-old obese male over 400 pounds. His scale maxes at 350, so he was guessing he probably weighted around 400 pounds. He presented with a lot of work-related stress. Having been remarried now for three years, he had teenagers in the home including step kids and biological children. He recognizes that at 56 years of age unless he loses about half his body weight, his weight is going to kill him.

In fact, he was actually surprised that he was still alive. He had come to me following a series of tests because he was considering bariatric surgery. For whatever medical reason though, he was ruled out as a candidate for bariatric surgery, so he came to me stressed, obese, and depressed—catastrophically depressed, obese to the point where he was at risk of imminent death, and stressed out from remarriage, children, work-related obligations and other things in life.

I worked with him for six sessions. (When I'm done talking about Bob, I'm going to share with you why Suzie was three sessions and why Bob was six sessions.) In my initial session with him, I had him identify his presenting problem, but also the strengths that he possessed. I inquired about his physical condition and his medication to rule out any health-related contraindications and of course to make certain that his physician knew that he was going to be addressing these issues with me. It's very important when we have somebody who is at such imminent risk of heart attack, stroke, or other health-related complications due to their physical condition that we make sure that they continue to work with a physician on these issues. And then I had him do a self-assessment of his strengths using a tool that I wrote called *The Nongard Strength and Resources Inventory.*

In that first session, I taught him Mindfulness. I remember he was one of those clients who said to me, "Okay, I can see how that would be helpful, but I'm not sure that's really going to change everything." And I said to him, "Well, you're right. Doing that right now here in my office probably didn't actually even change anything. But your assignment is to actually practice that exact same exercise each and every day, three times a day between now and your next session. In fact, I'm going to email an MP3 to your cell phone—that's what I do with my clients—and you're going to practice this with intention each and every day between now and our next session, which is going to be a week from now."

When he came in for session number two, and he said, "Wow, as far as my stress level, that Mindfulness exercise has been amazing. I did it for the first two or three days, and I thought to myself, 'Oh, I can see how this would be helpful,' but about halfway into the week, I began to find myself sitting in my car or even at my desk and being mindful rather than stressed out. That's been really helpful."

To extend his Mindfulness practice, we did the Raisin Exercise from Chapter Eight.

Now, one of the things that was interesting about him is that he was also an adjunct university professor, and he loved teaching at that level. He taught one class a semester, but he'd been doing that for more than twenty years, and he really enjoyed contributing to his field by educating young people. So I actually took my dry-erase marker and my marker board, and I did a fifteen-minute lecture on Relational Frame Theory for him, teaching him what I taught you in our chapter on relational frames. Then I spent some time with him looking at some of the frames that he'd created. Next I asked him to practice breaking relational frames with Mindfulness, both in the session and over the next week.

I used the Miracle Question during session number three. I often ask the Miracle Question during the first session, but I waited until the third session with him because I wanted to establish some goals related to his work stress and related to his marriage. Also, I wanted to develop some rapport with him before I really began a questioning process. I could tell just by his personality that teaching him some skills that he could immediately apply was going to be agenda number one and then going back for a questioning approach would be something that would be more effective after developing therapeutic rapport.

Maybe it's because I'm an educator, maybe it's because I'm an personal fitness trainer, or maybe it's because I recognize that people only do what they know how to do—in any case, I spent part of session three teaching him again. I taught him, not a complete course in nutrition and multiple food choices, but a couple of healthy food choices that he could make.

I often ask my clients, "What is it that you are eating? What do you have for lunch today? What did you have for dinner last night?"

People always think that what they're eating is healthy, but rarely is it, so I can help them with their food choices. With him, I also gave him some ideas for preparing for those days when he was going to be unable to eat at home and was going to have to eat out.

I also gave him some shopping instructions. This is something I teach to a lot of people whom I see for bariatric counseling. A lot of the clients whom I work with shop like Americans, not like Europeans; if you want to be healthy, you'll shop like a European or an Asian rather than an American.

In America, we usually get the Thursday newspaper. We cut out all the coupons. We wait until Friday payday and then we take our cart to the grocery store that always has everything and is never out of anything, and we load up two carts with all of the stuff that we can possibly use our Thursday coupons for with our Friday paycheck. Then we buy it all and put it in the deep freezer, and we eat that over the next two weeks before we repeat the process with new coupons and a new paycheck. When people shop like that, they're purchasing unhealthy foods: prepackaged, frozen foods that can be stored rather than food that is fresh and healthy.

I spend a fair amount of time in Eastern Europe and in Asia as well, and I have noticed something really interesting there. Never once did we go to the grocery store and buy food for a week.

Instead, everyday at the end of the day, we would go to the market before dinner and buy everything we needed for dinner tonight and that which we needed for breakfast tomorrow. What that meant is we were always buying fresh food. We were going to be eating the food tonight that we were buying right now and only storing food until the morning because we were going back to the market again tomorrow.

(If you watch British television, you may notice that characters coming home are always carrying a bag of groceries. That's because they shop and eat European-style, not American.)

I tell clients to shop daily for groceries rather than weekly for groceries. It makes a huge difference in the quality of foods that they will be eating.

In session number four with Bob, I reviewed how his Mindfulness practice had resulted in automatic mindful awareness, and he reported that his stress level was dramatically decreased at work. I used a Likert scale, and he had gone from an 8 or a 9 in week one down to a 3 or 4. In regards to his marriage and his home life, although he still found it stressful at a level 6 or a 7, when he came to see me he said it was an 8 or 9 and heading for a 9 or a 10. So that's an improvement of several points.

He found that this course of treatment to be so useful to him that he wanted to share it with somebody. So, in session number five, he brought his spouse not to fix her, but because he wanted her to hear from me the things that I had taught him that he thought would be helpful to her. In other words, he didn't bring her saying, "You know, my wife is screwed up, and she needs a therapist, too." He brought her because he wanted to say, "Honey, wow this guy has really helped me by teaching me some things, and maybe they're useful to you, too." So in session number five,

I really went back to lesson number one because this was her starting point and for him it was a great review.

In session number six, we talked more about food preparation, food choices, and other practical things related to weight loss. My clients do not go on a diet. What they do is they make substantial changes in their food choices, the way that they prepare food, and the way that they eat their food. I generally follow the recommendations of a nutritionist lifestyle.

I'm not going to call it a diet because it's not something my clients are ever going to end. There are a lot of great books in this area. Some of you may be familiar with John McDougall's books. Those are something that I certainly recommend. My favorite book to recommend is one that I actually keep in my office and give to all of my bariatric clients: Joel Fuhrman's *Eat to Live*. I buy copies ten or twenty at a time, stack them in the bottom shelf of my bookcase in the office, and give those away to clients all the time.

In session number six, he decided to follow up our six sessions by participating in our eight-week online Mindfulness Meditation Development Program.

After our initial six sessions, I probably saw him for four or five more sessions on a monthly or quarterly basis. When I have clients who are that big, I want to make sure that they're staying on track. In the past year and a half, he has come very close to the weight loss goals which he had established for himself, and he is continuing to improve.

Notes on the Case Studies

With neither one of these clients did I try to find out why they were anxious or obese. I simply dealt with that which they could change: the present. Both clients found they could change the present by applying the ideas of Mindfulness, Acceptance and the other aspects of Contextual Psychology discussed in this book.

Now, I mentioned that I would share with you a structure of three or six sessions. I offer my services as brief therapy. I want my clients to know that therapy doesn't last forever. I let my clients know that there may be situations and scenarios where they might want to access my services in the future. For my bariatric clients who are losing so much weight, we might meet on a quarterly basis. But I typically schedule my clients and have them commit to either three or six session packages.

The reason is as important as I want my clients to be invested in the process, so I want them to commit depending on the nature of the presenting problem to three or to six sessions. I do a lot of work with cigarette smokers, and I see them for three sessions. When I see somebody for pre-surgical anxiety, three sessions is a great number. When I'm dealing with somebody like severe obesity, six sessions and additional follow ups are probably useful.

If I have a client who has a lifelong history of catastrophic depression or bipolar disorder, probably six sessions to really make sure that they can focus on these things between sessions is important.

Bibliography

"Internet Encyclopedia of Philosophy", University of
     Tennessee http://www.iep.utm.edu/freud/ (accessed
     Mardh 8, 2014 2014).

, Association of Contextual Behavioral Science http://
     contextualscience.org/contextual_psychology (accessed
     March 8, 2014 2014).

"The Six Core Processes of Act", Association for Contextual
     Behavioral Science http://contextualscience.org/the_six_
     core_processes_of_act (accessed March 8 2014).

"What Is Mindfulness-Based Stress Reduction?" http://
     www.mindfullivingprograms.com/whatMBSR.php (accessed
     March 8 2014).

Atkinson, William Walker and Harry Houdini Collection
     (Library of Congress). *Thought Vibration, or, the Law
     of Attraction in the Thought World*. Chicago, U.S.A.:
     Library Shelf, 1910.

Baer, Ruth A. *Mindfulness-Based Treatment Approaches:
     Clinician's Guide to Evidence Base and Applications*
     Practical Resources for the Mental Health
     Professional. Amsterdam ; Boston: Elsevier,
     Academic Press, 2006.

Barber, T. X. "Hypnotic Age Regression: A Critical Review." *Psychosom Med* 24, (1962): 286-99.

Baumeister, Roy F. and John Tierney. *Willpower : Rediscovering the Greatest Human Strength*. New York: Penguin Press, 2011.

Bishop, Scott, Mark Lau, Shauna Shapiro, Linda Carlson, Nicole Anderson, James Carmody, Zindel Segal, Susan Abbey, Michael Speca, Drew Velting and Gerald Devins. "Mindfulness: A Proposed Operational Definition." *Clinical Psychology: Science and Practice* 11, no. 3 (2004): 230-241.

Bond, C., K. Woods, N. Humphrey, W. Symes and L. Green. "Practitioner Review: The Effectiveness of Solution Focused Brief Therapy with Children and Families: A Systematic and Critical Evaluation of the Literature from 1990-2010." *J Child Psychol Psychiatry* 54, no. 7 (2013): 707-23.

Davis, D. M. and J. A. Hayes. "What Are the Benefits of Mindfulness? A Practice Review of Psychotherapy-Related Research." *Psychotherapy (Chic)* 48, no. 2 (2011): 198-208.

De Shazer, Steve, Yvonne M. Dolan and Harry Korman. *More Than Miracles : The State of the Art of Solution-Focused Brief Therapy* Haworth Brief Therapy Series. New York: Haworth Press, 2007.

Egan, Gerard. *The Skilled Helper : A Problem-Management and Opportunity-Development Approach to Helping*. 10th ed. Belmont, Calif.: Brooks/Cole, Cengage Learning, 2014.

Fredrickson, B. L. and M. F. Losada. "Positive Affect and the Complex Dynamics of Human Flourishing." *Am Psychol* 60, no. 7 (2005): 678-86.

Freud, Sigmund, Josef Breuer and Nicola Luckhurst. *Studies in Hysteria*. London ; New York: Penguin Books, 2004.

Haanel, Charles F. *The Master Key System in Twenty-Four Parts with Questionnaire and Glossary*. Saint Louis, Mo., 1919.

Hayes, Steven C. *Get out of Your Mind and into Your Life*. New York, NY: MJF Books, 2011.

Hayes, Steven C., Kirk Strosahl and Kelly G. Wilson. *Acceptance and Commitment Therapy : The Process and Practice of Mindful Change*. 2nd ed. New York: Guilford Press, 2012.

Ievleva, L. and T. Orlick. "Mental Links to Enhanced Healing: An Exploratory Study." *Sports Psychologist* 5, no. 1 (1991): 25-40.

James Carson, Kimberly Carson, Karen Gil and Donald Baucom, "Mindfulness-Based Relationship Enhancement", Association for Advancement of Behorior Therapy http://www.bemindful.org/mbrelenhanc.pdf (accessed March 8 2014).

Janov, Arthur. *The Primal Scream; Primal Therapy: The Cure for Neurosis*. New York,: Putnam, 1970.

Kabat-Zinn, Jon. *Wherever You Go, There You Are : Mindfulness Meditation in Everyday Life*. 1st ed. New York: Hyperion, 1994.

Linehan, Marsha. *Skills Training Manual for Treating Borderline Personality Disorder* Diagnosis and Treatment of Mental Disorders. New York: Guilford Press, 1993.

Metcalf, Linda. *The Miracle Question : Answer It and Change Your Life*. 1st Ed. ed. Williston, VT: Crown House Pub., 2007.

Miller, William R. *Integrating Spirituality into Treatment : Resources for Practitioners*. 1st ed. Washington, DC: American Psychological Association, 1999.

Nongard, Richard. *Medical Meditation: How to Reduce Pain, Decrease Complications and Recover Faster from Surgery, Disease and Illness*. Scottsdale, AZ: Peachtree Professional Education, Inc., 2011.

Satir, Virginia. *The Satir Model : Family Therapy and Beyond*. Palo Alto, Calif.: Science and Behavior Books, 1991.

Seligman, M. E. and M. Csikszentmihalyi. "Positive Psychology. An Introduction." *Am Psychol* 55, no. 1 (2000): 5-14.

Seligman, M. E., T. A. Steen, N. Park and C. Peterson. "Positive Psychology Progress: Empirical Validation of Interventions." *Am Psychol* 60, no. 5 (2005): 410-21.

Swenson, C. R., C. Sanderson, R. A. Dulit and M. M. Linehan. "The Application of Dialectical Behavior Therapy for Patients with Borderline Personality Disorder on Inpatient Units." *Psychiatr Q* 72, no. 4 (2001): 307-24.

Teasdale, J. D., Z. V. Segal, J. M. Williams, V. A. Ridgeway, J. M. Soulsby and M. A. Lau. "Prevention of Relapse/Recurrence in Major Depression by Mindfulness-Based Cognitive Therapy." *J Consult Clin Psychol* 68, no. 4 (2000): 615-23.

Toma, C. L. and J. T. Hancock. "Self-Affirmation Underlies Facebook Use." *Pers Soc Psychol Bull* 39, no. 3 (2013): 321-31.

Watkins, Jane Magruder and Bernard J. Mohr. *Appreciative Inquiry : Change at the Speed of Imagination* Practicing Organization Development Series. San Francisco, Calif.: Jossey-Bass/Pfeiffer, 2001.

Wood, J. V., W. Q. Perunovic and J. W. Lee. "Positive Self-Statements: Power for Some, Peril for Others." *Psychol Sci* 20, no. 7 (2009): 860-6.

www.ingramcontent.com/pod-product-compliance
Lightning Source LLC
Chambersburg PA
CBHW032012170526

45157CB00002B/664